PLASTIC CONVERSION PROCESSES

A Concise and Applied Guide

Eric Cybulski

PLASTIC CONVERSION PROCESSES
A Concise and Applied Guide

CRC Press
Taylor & Francis Group
Boca Raton   London   New York

CRC Press is an imprint of the
Taylor & Francis Group, an **informa** business

CRC Press
Taylor & Francis Group
6000 Broken Sound Parkway NW, Suite 300
Boca Raton, FL 33487-2742

© 2009 by Taylor and Francis Group, LLC
CRC Press is an imprint of Taylor & Francis Group, an Informa business

Printed in the United States of America on acid-free paper
10 9 8 7 6 5 4 3 2 1

International Standard Book Number: 978-1-4200-9406-0 (Paperback)

Library of Congress Cataloging-in-Publication Data

Cybulski, Eric.
  Plastic conversion processes : a concise and applied guide / Eric Cybulski.
    p. cm.
  Includes bibliographical references and index.
  ISBN 978-1-4200-9406-0 (pbk. : alk. paper)
  1. Plastics. I. Title.

TP1120.C93 2009
668.4--dc22

2009014840

Visit the Taylor & Francis Web site at
http://www.taylorandfrancis.com

and the CRC Press Web site at
http://www.crcpress.com

Contents

Preface

During the plastic material development explosion, which took place in the late 20th century, a proliferation of conversion processes and their variants arose. Although books are available describing a single process like injection molding, sources that describe and compare a comprehensive list of plastic conversion processes do not exist.

This book provides a basic overview of seven conversion processes used in the industry. These processes account for more than 97% of all plastic products. Each chapter begins with a process attribute table to serve as a quick guide. The particular conversion process is then briefly described along with a short history. To understand the process better, sections detailing equipment, tooling, and materials have been added. Finally, a general design guide and case studies complete each section. As an added bonus, more than 350 terms and definitions are included in Appendix A: Plastics Terms, Definitions, and Examples from A to Z.

This book was written to allow the comparison, evaluation, and selection of the best process for your product. It is easy to understand and supplemented with diagrams and pictures. The intent is to characterize the industry in a manner that is not intimidating and to acquaint those new to the field with their possible choices.

Acknowledgments

I am grateful to the many people who supported me while I developed this book. Some were technical reviewers, others helped guide my writing style. It takes incredible energy and attention to detail. I am thankful to those who took their time to do it.

Claude Cybulski—3M

Chris Eriksen

Michael D. Johnson—Texas A&M University

Debbie Judd

MGS Mfg. Group—Germantown, WI

Climatech—Hopkins, MN

Special thanks to my father for introducing me to the world of engineering and plastics, my mother for being supportive, and my wife, Sandi, for encouraging me to write this book.

Eric Cybulski

About the Author

A desire to build and design products led Eric Cybulski to a career in engineering

Eric graduated from the University of Minnesota with a degree in mechanical and industrial engineering. He has held plastic product development and commercialization positions for more than 11 years in both the private and public sectors, accounting for revenues well over $1 billion. His unique understanding of plastic product development and conversion processes has resulted in the filing of 25 U.S. patents to date.

He and his wife enjoy spending time with family and friends.

1

Injection Molding

Injection Molding Process Key Characteristics

Volume	High
Material selection	Extensive
Part cost	Very low
Part geometry	Complex
Part size	Very small to large
Tool cost	Very high
Cycle time	Seconds
Labor	Automatic

For a list of other conversion process characteristics, see Appendix B.

1.1 Process Overview

Injection molding is the most common and versatile of all plastic conversion processes. In this process, plastic pellets, also known as resin, are fed through a barrel with a series of heaters that aid in melting the resin pellets and maintaining the barrel temperature. The material is then injected under high pressure into the mold cavity. Once the plastic has cooled, the mold opens to eject the part.

Injection molding is capable of a wide range of part sizes and complex geometries. A molding cycle consists of five basic stages: fill, pack, hold, cool, and eject. In the fill stage, the B-side of the mold closes on the A-side of the mold with enough clamping force to prevent flashing as molten plastic is injected into the mold cavities through the gate. At this time, the cavities are 95%–98% full. Pack is the second stage of the injection molding cycle. It fills the mold, volumetrically, before the press transfers over to a holding pressure to allow the gate to freeze. Once the gate freezes off, the screw retracts and prepares for the next cycle. If the gate is not allowed to freeze off and the screw rotates, some material will flow back from the mold into the barrel. Finally, after the part has cooled enough, the mold opens and the part is ejected. Figure 1.1A illustrates a diagram of an injection molding press. These presses are available in a wide range of sizes, clamp tonnages, and shot volumes as shown in Figures 1.1B through 1.1D.

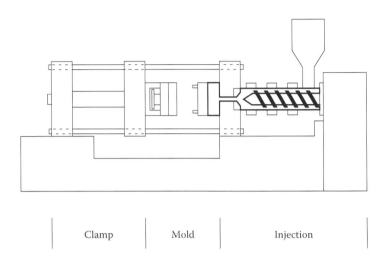

| Clamp | Mold | Injection |

FIGURE 1.1A
Diagram of an injection molding press.

FIGURE 1.1B
Injection molding press. *Source:* Photo courtesy of MGS Mfg. Group.

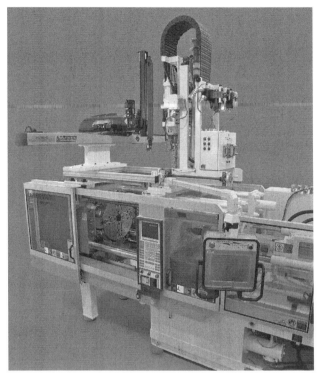

FIGURE 1.1C
Injection molding press. *Source:* Photo courtesy of MGS Mfg. Group.

FIGURE 1.1D
Injection molding press. *Source:* Photo courtesy of MGS Mfg. Group.

Why would you use this process? It is well suited for high volume, complex parts that have tight tolerances.

The most familiar applications of this process include

- Telecommunications
- Medical and pharmaceutical
- Consumer and office products
- Electronics
- Automotive
- Toys
- Containers and closures
- Plumbing
- Packaging

1.1.1 Variations of the Injection Molding Process

There are three types of injection molding presses in use today: electric, hydraulic, and hybrid. The hydraulic press has superior clamping pressure and accounts for a majority of the injection molding presses used in the industry. Electric presses are faster, more accurate, more environmentally friendly, and quieter than their counterpart, the hydraulic press. Due to their inherent advantages, electric presses are going to continue to increase in popularity. Lastly, hybrid presses combine features from both the electric press and the hydraulic press, which includes the clamping pressure of a hydraulic press along with the accuracy and energy efficiency of an electric press.[1] Molding process variations include multi-shot, gas and water assist, and structural foam.

Multi-shot injection molding is a process where two or more materials are molded within a single cycle. Figure 1.2 shows an example of a multi-shot rotary platen injection molding setup. The vast majority of multi-shot parts involve a thermoplastic material and an elastomer. Another form of the multi-shot molding is spin stack molding. The first material is molded and the entire mold rotates to the next stage where the second material is molded. The most recent advancement in multi-shot molding is referred to as "in-mold assembly." This is a method of molding multiple parts capable of independent movement in one cycle. Examples include valves, air vents, and locking mechanisms. Multi-shot may also be used to create soft-touch surfaces on products.

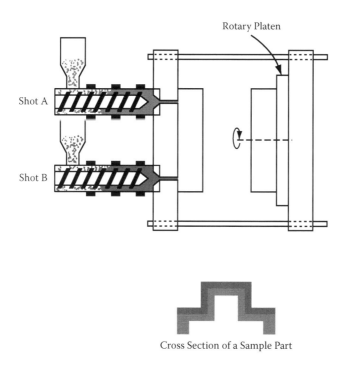

Cross Section of a Sample Part

FIGURE 1.2
Multi-shot injection molding setup.

Gas assist is a method in which a compressed gas, typically nitrogen, is injected into the center of the melt stream as shown in Figure 1.3 to displace material by reducing the effective wall section. The pressure is maintained until the part cools and the gas is evacuated from the part. This process uses less injection pressure, which translates into less clamping pressure. Gas assist injection molding has been used to minimize sink and warp in thick or nonuniform wall section parts. Parts having a high dimensional aspect ratio, uniform wall section, and thicknesses exceeding 0.250″ (6.35 mm) are candidates for gas assist molding.

The structural foam method is used when molding parts with cross sections larger than 0.175″ (4.45 mm). Material is either pre-blended with a blowing agent or an inert gas is introduced directly into the melt stream. As the material enters the mold cavity, small gas bubbles expand and fill out the part. Figure 1.4 shows structural foam equipment. Gas bubbles will form drag marks along the surfaces of the parts, so a smooth appearance is not possible using structural foam molding. Pressurizing the mold helps to minimize but not eliminate surface bubbles and improves appearance. This

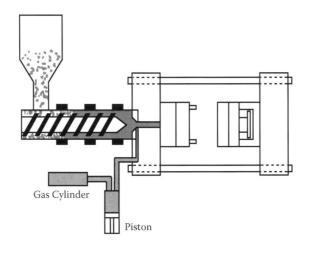

Cross Section of a Sample Part

FIGURE 1.3
Gas assist molding equipment.

is referred to as applying counter pressure to the mold. Similar to gas assist injection molding, this method of molding utilizes considerably less clamping pressure and less cavity pressure than conventional injection molding.

1.2 A Brief History of Injection Molding

The concept of plastic injection molding was first developed by John Wesley Hyatt in 1868, in response to a challenge from a company searching for an alternative material for ivory. At the time, ivory was primarily used for billiard balls and piano keys, but was becoming increasingly hard to find. A reward of $10,000 was offered for the solution. Hyatt accepted the challenge and using a plunger-style injection molding machine, he molded billiard balls from celluloid, which was a mixture of cellulose nitrate and camphor. Four years later, he and his brother Isaiah patented the plunger-style injection molding machine, and for the next 50 years not much changed in the

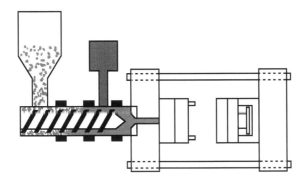

Cross Section of a Sample Part

FIGURE 1.4
Structural foam molding equipment.

how plastics were molded, but numerous discoveries were made in the materials area of molding.

The first materials available for processing were thermosets. In 1907, Bakelite was invented by Leo Baekeland. Polystyrene was first developed commercially by BASF in 1930 and later in the United States by Dow Chemical in 1937.[2] In 1933, Eric Fawcett and Reginald Gibson of Imperial Chemical Industries invented the first thermoplastic, polyethylene.[3] During this time, a few new materials were introduced as a number of well known companies began efforts in plastics research and development. These companies included, but were not limited to, B. F. Goodrich (polyvinyl chloride [PVC]—U.S. patents 1,929,453 and 2,188,396), General Electric, Dow Chemical, DuPont, and BASF. As tension mounted among European countries and Asia was in turmoil, these materials and others would have an impact on the outcome of World War II. As trade relations with Japan rapidly deteriorated, DuPont began searching for an alternative material for silk when nylon was invented.

With World War II involving countries worldwide, the demand for plastics skyrocketed as metal and other common materials were earmarked to support the war effort. Items that were injection molded during the war for military use included canteen caps, MP38 machine gun components, radio enclosures, buttons, eating utensils, and even components for the atomic bomb.[4][6]

The next significant innovation to occur in the field of injection molding was the introduction of the first screw injection molding machine, which was patented by James Hendry in 1946.[7] Shortly afterward, the reciprocating screw was invented in 1952 and patented in 1956 by William Willert. Modern-day injection molding was born. Many consider the reciprocating screw the single most important contribution that revolutionized the plastics industry in the 20th century.[8]

Injection molding processes and equipment continue to change. For example, today multi-shot injection molding is utilized to fabricate an ever-increasing number of parts. It is most commonly used with an elastomer to provide the familiar "soft touch" part. The first material is molded and allowed to cool. It then transfers to a second position where the next material is molded over the previous material. This can be repeated multiple times. In-mold assemblies are starting to become more common as well.

1.3 Equipment

Equipment used in the injection molding process is listed below, and Figure 1.5 illustrates the injection molding press areas. Basic descriptions of the injection molding press components appear on the following pages to provide a better understanding of their function and purpose.

1 – Hopper
2 – Barrel
3 – Reciprocating screw
4 – Heat regions
5 – Nozzle
6 – Stationary platen
7 – Mold - Standard two-plate mold components
8 – Moving platen
9 – Tie bars

1.3.1 Hopper

The hopper, also used on extrusion and blow molding equipment, funnels unmelted plastic pellets, by gravity, to the feed section of the barrel. Some hoppers will have a transparent window to view the material level. Material can be added manually or with an attached vacuum system for high throughput applications. Hoppers are covered to prevent possible contamination and also feature a magnetic screen above the entrance to the throat

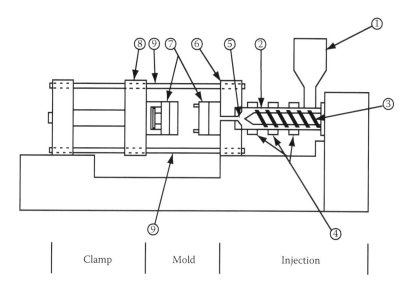

FIGURE 1.5
Injection molding equipment diagram.

of the barrel to catch any metal fines, chips, bolts, or other small objects that may be accidentally dropped into the hopper. Metal contaminants can seriously damage the screw.

1.3.2 Barrel

The purpose of the barrel is to house the reciprocating screw and provide the material delivery path to the mold. The main components of a barrel include the barrel sheath, the reciprocating screw (also referred to as the screw), a hydraulic system for moving the screw forward and backward, and a series of heater bands, which are used to maintain the proper melt temperature of the material.

1.3.3 Reciprocating Screw

The screw is designed so that when it rotates, the resin pellets are metered forward by the flights, gradually melting and building up pressure along the way. Typical clearance between the flights of the screw and the barrel wall ranges from 0.003″ to 0.010″ (0.076 mm to 0.254 mm), depending on the material being processed. The depth of the flights, which is the distance from the outer edge of the flight to the shaft of the screw, varies depending upon the section of the screw. The screw is divided into three sections or zones: feed zone, transition zone, and metering zone as shown in Figure 1.6.

In the feed zone, the screw has the largest flight depth so the unmelted plastic pellets can enter the barrel via the feed throat and be moved forward

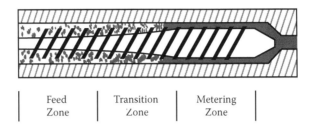

| Feed | Transition | Metering |
| Zone | Zone | Zone |

FIGURE 1.6
Reciprocating screw that features three zones.

to the next zone. As the pellets enter the transition zone, the depth of the flights gradually decreases, which in turn increases the shear and pressure of the resin against the screw flights. This mechanical action melts the pellets, which helps to reduce any imperfections in the feed, eliminates entrapped air pockets, and ensures a homogeneous resin melt. The resin then enters the metering zone, where the flight depth is at its minimum. Finally, the melt passes through a check ring at the tip of the screw. At this point, the screw stops rotating and becomes a ram moving forward to inject the polymer into the mold. It repeats this reciprocating action for every cycle.

1.3.4 Heat Regions

The heat regions, controlled by heater bands, maintain a constant temperature of the material in the barrel within a zone. They are not the primary sources of heat generation; the majority of the heat is shear heat created from the friction generated by the compression of the plastic pellets in the barrel by the screw. In most cases, injection molding press barrels have three or more independently controlled heater band regions to help maintain the desired temperature of the material being extruded through the barrel.

1.3.5 Nozzle

The nozzle provides the interface between the extruder and the mold. It is positioned at the end of the barrel and is aligned to the sprue bushing hole in the mold. A small heater band is attached to the nozzle to compensate for heat loss effects when it is in contact with the mold.

1.3.6 Stationary Platen

The A-side of the mold is mounted to the stationary platen, which is located near the nozzle of the injection molding press. It does not physically move during the process, but serves as the surface against which the press clamp exerts pressure.

1.3.7 Mold

An isometric view of an injection mold is illustrated in Figure 1.7. The mold is divided into two halves, the A-side (exploded view shown in Figure 1.8) and the B-side (exploded view shown in Figure 1.9). A description of a basic two-plate mold follows. The A-side consists of the top clamp plate, the A-plate, the sprue bushing, the locator ring, and leader pins. The sprue bushing and the locator ring are bolted to the top clamp plate and assist in aligning the mold to the nozzle of the press. Four leader pins are inserted through the A-plate and held in place by the top clamp plate to ensure proper alignment of the mold halves as it closes. This minimizes part detail mismatch between the mold halves. Finally, the A-plate contains a portion of the molded part's details.

The B-side consists of the B-plate, the support plate, the bottom clamp plate, and the ejector assembly (exploded view shown in Figure 1.10) all bolted

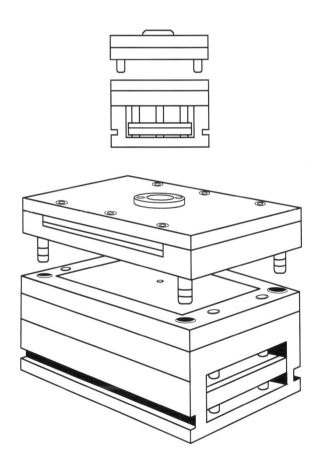

FIGURE 1.7
Isometric view of an injection mold.

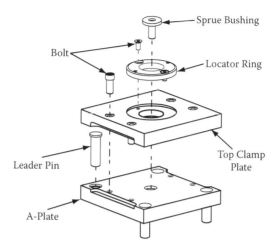

FIGURE 1.8
Exploded view of the A-side of an injection mold.

together like the A-side. Typically, the B-side cavity contains the inside part detail. The support plate is attached to the B-plate and the bottom clamp plate. The support plate helps to minimize the deflection of the B-plate during high-pressure injection. Support pillars are also attached to the bottom clamp plate and serve to add additional strength to the support plate. The ejector assembly moves forward within the mold to assist in the ejection of the part by means of ejector pins and returns to its home position by means of a set of return pins.

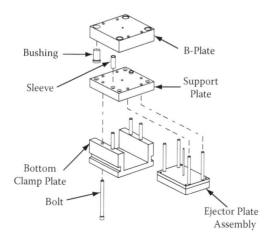

FIGURE 1.9
Exploded view of the B-side of an injection mold.

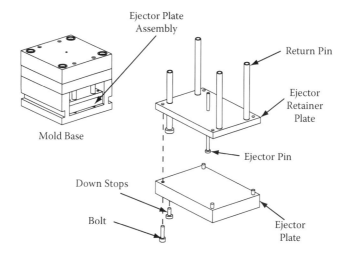

FIGURE 1.10
Exploded view of the ejector plate assembly of an injection mold.

1.3.8 Moving Platen

The B-side of the mold is mounted to the moving platen. The return pins on the moving platen are attached to the ejector assembly to ensure that the ejector pins are retracted prior to the mold closing. Once the molding cycle begins the B-side platen advances toward the stationary platen. This side of the press is also responsible for clamping against the A-side of the mold during the injection stage.

1.3.9 Tie Bars

The tie bars serve to guide the press platens on the injection molding machine and allow the press to develop the required clamping force. The spacing between the tie bars defines the maximum space allowed for the mounting and removal of a mold from the press. This also defines the maximum mold size.

1.4 Tooling

It is not the intent of this book to explain tool construction and mold design in great detail; other sources exist. An overview is presented here to make the reader aware of the complexity of the tooling involved in this process. It is important to understand the impact of each process on the outcome of the product. Integral to a successful program is the proper execution of part design, tooling design, and process management.

1.4.1 Mold Materials

The mold material selected to fabricate the mold is just as important as the resin selected for molding the part. Injection molds are primarily fabricated from aluminum and steel. These materials have the strength to withstand high injection pressure, dissipate heat, and provide adequate wear character-istics. Aluminum dissipates heat better than steel, but wears much quicker. Various types of tooling steel are also used, which include H-13, P-20, and S-7. Stainless steel can be used in some cases. P-20 is the most common mate-rial used for fabricating medium run injection molds. Aluminum is typically used for prototype tooling and lower volume parts.

1.4.2 Mold Fabrication

One highly accurate mold fabrication process is called electrical discharge machining (EDM). Accuracies of ± 0.0001″ (± 0.0025 mm) are achieved using the EDM process. EDM uses an electrode, which is machined from graphite or a metal alloy to burn engineered geometry into the tool mate-rial. Graphite is the preferred material for electrodes because it is easy to machine, has a high melting temperature, and has a high ratio of material removal to wear. A current is applied to the electrode and a spark is pro-duced between the electrode and the work piece, which causes the metal to melt away. The higher the current, the more material will be melted away. Using the EDM process, fine complex details can be created in the mold.

Molds can feature either a cold runner, which is ejected after each cycle, or a hot runner, which does not solidify after each cycle. Hot runner molds are more expensive than cold runner tools, but will cycle faster, thus recover-ing the tool cost in high volume product applications. Often a heated sprue bushing is used with a cold runner system to decrease the cooling time by not waiting for the sprue to solidify. The thickness of the sprue can dictate the cycle time.

Since cooling is 80% of the cycle, water lines are added to the mold to aid in the removal of heat from the melted plastic. The more turbulent the cooling line flow, the higher the heat transfer rate, which means a better cooling rate. The design of the mold should use a water line of 0.43″–0.50″ (10.9 – 12.7 mm) in diameter, uniformly placed around the cavity and approximately 0.50″ (12.7 mm) from the cavity surface.

Injection molds have very tight tolerances between components, so during assembly proper orientation of the plates is critical. An identification mark, which appears as a "0", is stamped into one corner of each plate to assist in plate alignment and orientation during assembly and maintenance work. To assist tooling makers with the separation of the mold plates, ply bar slots are machined into the corners of the plates so they can be pried apart easily.

Gate design is an important aspect of tool design. The primary gate designs used by mold makers include edge gates, fan gates, flash gates, sprue gates, tab gates, tunnel gates, and valve gates. When multiple gates are used to mold a part, a knit line will be present in the part where the two flow fronts meet. This area can be weak and subject to fracture, depending on the part design.

1.5 Materials

There is an extensive list of materials available for injection molding; more material grades are available for injection molding than all other processes combined. Materials used in this process are delivered to the molder in the form of pellets. The resin processed is either natural or compounded. Natural resin does not have any additives or fillers. Compounded resins include a wide variety of material additives and fillers. Examples include colorants, glass fibers, impact modifiers, talc, ultraviolet inhibitors, or wood. These are added to the base resin to enhance one or more properties. Additives and fillers can be compounded into the resin by some molders right at the press. Table 1.1 lists some of the more common materials used in the injection molding process along with their properties and some of the main applications.

1.6 Injection Molding Part Design Guidelines

Injection molding is the most common of all plastic conversion processes, but designing parts for this process can be the most difficult. It is well suited for complex parts as well as simple geometries. Following these basic guidelines can help to eliminate problems down the road. It is important to check with the molder for the exact design guidelines based on the part to be molded, as this can vary from molder to molder. The information that follows provides some basic guidelines.

Wall sections (nominal; where t is equal to the nominal wall thickness)
- Maintain an optimum uniform wall thickness ranging from 0.010″–0.157″ (0.25–4.0 mm).
- Minimum of a 1° draft angle [1° = 0.017″ per inch (1° = 0.43 mm per mm)] required to eject the part.
- Maintain a gradual material transition without sharp edges to reduce molded-in stress as shown in Figure 1.11.
- Boss draft of 0.25° possible using a sleeve ejector.

TABLE 1.1

Common Injection Molding Materials

Material	Properties	Sensitivity to Moisture	Applications
ABS	High impact Weather resistant Rigid	Yes	Tool housings Remote controls Plumbing Impact applications
HIPS/PS	Low process temperature High impact	No	Packaging Toys Impact applications Low cost applications
PC	Rigid High impact Excellent clarity Heat resistant	Yes	Automotive headlights Fixtures Medical devices
PE—HDPE and LDPE	Chemical resistant Translucent Good impact High shrinkage rate	No	Low cost applications Medical applicators Handles Drinking water devices
Polyamide	Good impact High process temperature Highly chemical resistant	Yes	High temperature applications Automotive components Chemical applications
PP	Good impact High process temperature Highly chemical resistant	No	Enclosures Chemical applications Food storage containers Toys
PVC	Good impact Highly chemical resistant Rigid	No	Plumbing Chemical applications
SAN	Transparent Brittle	No	Drinking water devices Viewing windows

Texture

- Draft texture 1° per 0.001″ (0.025 mm) of texture depth.

Radii

- Minimum radius of 0.5 t of the nominal wall thickness as shown in Figure 1.12.
- Minimum radius of 0.03″ (0.08 mm) with a 0.06″ (1.5 mm) radius preferred for wall sections less than or equal to 0.06″ (1.5 mm).
- The larger the radius the better to minimize stress concentrations.
- No sharp edges.

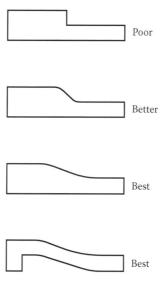

Poor

Better

Best

Best

FIGURE 1.11
Design guide—illustration showing wall transition variation.

Ribs

- Single rib designs are shown in Figure 1.13—depending on the material processed, ribs should be a maximum of 0.4 t–0.6 t of the mating wall section or sink may appear on the part.
- Ribs exceeding 0.4 t–0.6 t should be broken down into multiple rib designs as shown in Figure 1.14. Maintain a 1 t–2 t spacing between ribs and a maximum height of 3 t.

Gussets

- Maximum wall thickness of .5 t–.7 t of the adjoining wall section, the thickness will depend on the material being processed.
- Maximum height of 3.5 t and a maximum length of 2 t.

Boss

- Maximum wall thickness of .4 t–.6 t of the adjoining wall section, the thickness will depend on the material being processed.
- Add gussets to increase the strength of the boss and ensure part fill.
- Holes will always have a knit line.

Holes

- No closer than 2 t from edges.
- Maintain a minimum spacing of 2 t between holes.

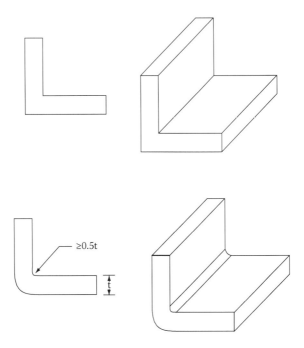

FIGURE 1.12
Design guide—illustration showing radii.

- Holes will always have a knit line.
- Minimum of a 1° draft angle, 2° preferred to assist in the de-molding process.

Living hinge

- Limited number of materials can be used to mold living hinges.
- Flex the part after it has been ejected from the mold to prolong the hinge life. Design guides specific to injection molded hinges are available.

Part attachment methods

- Snap fits—permanent snap designs and multiple use snap designs.
- Press fit and interference fit.
- Ultrasonic welding.
- Spin welding.
- Adhesive.
- Fasteners and inserts can be installed using ultrasonic welding.

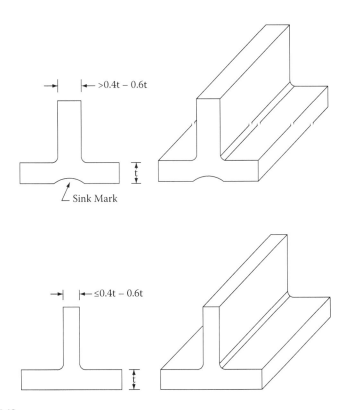

FIGURE 1.13
Design guide—illustration showing a single rib design.

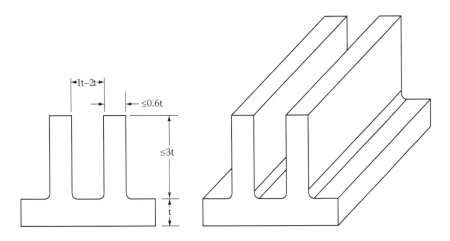

FIGURE 1.14
Design guide—illustration showing a multiple rib design.

1.7 How to Identify an Injection Molded Part

Below is a list of characteristics that can be used to assist in the identification of an injection molded part. These would include a gate, possible complex geometry, intricate snap fit features, or nonlinear parts with varying wall thicknesses.

- Complex geometry
- Living hinge
- Snap fits
- Texture
- Undercut
- Corners or sharp transitions
- Ribs
- Flowline or knitline
- Interference fits
- Bosses and posts
- Cavity identification

1.8 Case Studies

1.8.1 Case Study 1—Living Hinges (DVD Case/Video Game Case)

Living hinges are design features that can push plastic materials to their limits. A living hinge is a feature that allows the material to flex without breaking. In some cases, the hinge can be flexed in excess of one million times before failure. A familiar example of a living hinge is the DVD/video game case.

Multiple case designs are currently on the market, some featuring one living hinge and others featuring two living hinges. This allows the cases to be molded as one piece, whereas conventional jewel cases are typically made using three separate parts that snap fit together using a post and hole. These features were used instead of living hinges because of the material used to mold the parts. The material used for the one-piece DVD/video game case is also less expensive than the material used for the three separate piece jewel cases.

To prolong the life of the living hinge it is important to flex the hinge before it has completely cooled. Other features of a DVD/video game case include some sort of center snap retainer ring which is required to hold the disc in place when the case is closed. To ensure that the case remains closed another type of snap fit is used to secure the two sides in place. On the side opposite the center snap retainer ring can be some tabs. The purpose of the tabs is to hold the liner notes or in the case of video or computer games, the instruction manual. DVD cases are truly unique and complex parts that can only be molded by the injection molding process.

1.8.2 Case Study 2—Communication Device Housings

Plastic components used for cell phones have evolved over a number of years. Current designs are based on a series of older communication devices like handheld walkie-talkies, which were manufactured with metal exteriors. These devices had to be extremely rugged as they were used by the U.S. Army during World War II and subjected to Arctic colds as well as desert heat.

Knowledge gained from usage patterns and advancements in technology enabled small pocket-sized, less expensive models to be developed that were as durable as the metal walkie-talkies. Some time passed, however, before mobile phone technology was widely available to the public at prices that make them short-term conveniences instead of long-term investments.

References

1. Mikell Knights, "Electric, Hydraulic, or Hybrid?" Plastics Technology, http://www.ptonline.com (accessed October 8, 2008).
2. Mary Bellis, "Polystyrene and Styrofoam," Inventors, http://www.inventors.com (accessed September 7, 2008).
3. Tim A. Osswald, "Polyethylene: A Product of Brain and Brawn," Bags Inc., http://www.bags-inc.com (accessed May 20, 2008).
4. Olive-Drab, "Military Canteen," http://www.olive-drab.com (accessed July 26, 2008).
5. Nation Master, "Bakelite," http://www.nationmaster.com (accessed August 29, 2008).
6. Marshall Cavendish Corporation, *Inventors and Inventions* (New York: Marshall Cavendish, 2008).
7. International Plastic Laboratories and Services, "A History of Plastics," http://www.iplas.com (accessed August 1, 2008).
8. Editors of Plastic Technology, "50 Ideas That Changed Plastics," Plastics Technology, http://www.ptonline.com (accessed August 13, 2008).

2

Plastic Extrusion

Extrusion Process Key Characteristics

Volume	High
Material selection	Moderate
Part cost	Very low
Part geometry	Simple
Part size	Small to medium
Tool cost	Low
Cycle time	Seconds
Labor	Automatic

For a list of other conversion process characteristics, see Appendix B.

2.1 Process Overview

Plastic extrusion can be described as a continuous process whereby uniform profiles of indefinite length are extruded by melting plastic pellets in a heated barrel and forcing the material through the orifice of the shaping plate, commonly referred to as a die. The profile exits the die and continues to the cooling zone where the extruded form may pass through a series of sizing plates or forming tools to shape the profile to its final size. A water bath or compressed air jet is utilized to remove excess heat from the profile, which allows the part to stabilize dimensionally before handling. As the profile continues to cool, part marking or identification such as a date code, or a decorative wrap can be applied. Finally, the profile enters the cutter where it is cut to length and either wound up on a take-up roll or packaged. A diagram of the standard extrusion process is shown in Figure 2.1A. A standard extruder with a hopper and cooling trough are shown in Figures 2.1B and C, respectively.

Why would you use this process? It is well suited for high volume parts that feature a uniform cross section.

The most familiar applications of this process include

- Rigid pipes
- Flexible tubing
- Coated wires

23

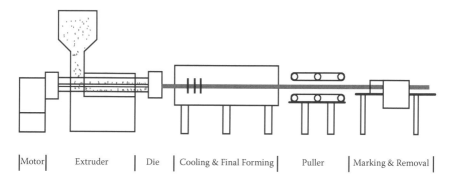

Motor | Extruder | Die | Cooling & Final Forming | Puller | Marking & Removal

FIGURE 2.1A
Diagram of an extruder.

FIGURE 2.1B
Extrusion equipment. *Source:* Photo courtesy of Climatech.

FIGURE 2.1C
Cooling trough. *Source:* Photo courtesy of Climatech.

- Building siding
- Window frames
- Weather stripping
- Decorative fencing

2.1.1 Variations of the Extrusion Process

There are three primary types of plastic extrusions: profile, coating, and sheet or film. This chapter covers primarily profile extrusion. Due to the high material throughput rate (pounds/hour), extrusion is one of the most cost-effective conversion methods available for high volume production; however, it is limited in applications.

A variation of the standard extrusion process is co-extrusion, shown in Figure 2.2, which is the process of extruding two or more materials simultaneously through a single die so that both materials are bonded together. Wire coating is another widely used extrusion variation, which is shown in Figure 2.3. Wire is fed into the melt stream and through the die. This process has been expanded into the window industry where a few specialized extruders feed wood instead of wire into the melt stream.

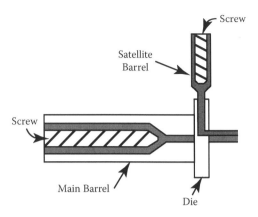

FIGURE 2.2
Co-extrusion process.

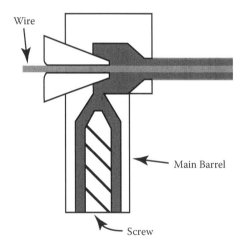

FIGURE 2.3
Wire coating process.

2.2 A Brief History of Extrusion

The first extruder was developed by H. Bewley of the Gutta-Percha Company in London, England, in 1847. The name gutta-percha comes from the genus of tropical trees that produce a sap, which is used to produce the inelastic natural latex that was later used as an insulation for coated wires. From 1848–1850, Charles Hancock, also of the Gutta-Percha Company, used H. Bewley's extrusion techniques to develop a wire coating, which was used for electrical insulation. This process produced the first original underwater telegraph cables.

In 1870, the first documented example of extrusion in the United States occurred when cellulose nitrate was extruded using a hydraulic ram as a means to deliver the material to the die.[1] This process was slow and limited to discrete lengths. The field of extrusion remained relatively unchanged until the early 1930s, after which there were significant advancements in both equipment and materials. New extruders featured a screw versus a hydraulic ram for the delivery system, which made extrusion a continuous process. In 1937, the first twin-screw extrusion machine to achieve high capacity was developed in Italy.[2] Concurrently, plastic materials were developed by companies such as Dow Chemical, DuPont, General Electric, Goodrich, and Imperial Chemical Industries. This research was accelerated during World War II, producing polyethylene film, insulation for wires (rubber was in short supply), and the groundwork for many new materials that would eventually be used in postwar products.

In a short time, American industry was converted from consumer goods to war production and back to consumer goods again. During the war, a considerable amount of money had to be invested in manufacturing by companies to meet the demands of the war, and now these lines were capable of mass-producing consumer goods. From the research conducted during World War II, polyethylene became one of the most popular plastics used in producing consumer goods. It became the first plastic material in history to sell over a billion pounds a year in the United States and it is currently the highest selling plastic by volume worldwide.[3] New materials and uses of extrusion began to appear in a wide variety of profile shapes and sizes, with one of the most famous of all time being the Hula Hoop®, made from polyethylene. It was first sold by Wham-O Inc. in 1958, and by 1960 sales of Hula Hoops in the United States exceeded 100 million, an achievement that no other toy has accomplished.[4] The Hula Hoop is considered one of the greatest fads in history.

Development in equipment, plastic materials, controls, die construction, screw design, and post process forming have moved extrusion from the art filled plunger process to an accurate high capacity commercial process. Extrusion can be a difficult process to control because the profile is actually formed to final dimensions after the die, versus injection molding or blow molding where the part is fully contained within the mold until it cools. Currently, extrusion plays an important part in the construction industry, supplying tubing, window frames, insulation strips, deck boards, and siding. It can be found in medical and automotive markets as well.

2.3 Equipment

Equipment used in the extrusion process is listed below and a sketch is shown in Figure 2.4, which breaks out the extruder into sections and the main individual components for a better understanding of their purpose and function. Basic descriptions of the main components of the extruder appear on the following pages.

1 – Hopper

2 – Barrel

3 – Extruder screw

4 – Thrust bearing

5 – Heat regions

6 – Breaker plates and screen pack

7 – Die

8 – Cooling zone

9 – Reshaping plates

10 – Puller

11 – Marking

12 – Cutter

2.3.1 Hopper

The hopper, which is also used on extrusion and blow molding equipment, funnels unmelted plastic pellets, by gravity, to the feed section of the barrel. Some hoppers will have a transparent window to view the material level. Material can be added manually or with an attached vacuum system for high throughput applications. Hoppers are covered to prevent possible

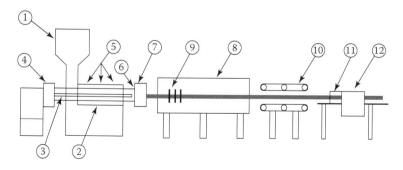

FIGURE 2.4
Extruder (with arrow callouts for each stage).

contamination and also feature a magnetic screen above the entrance to the throat of the barrel to catch any metal fines, chips, bolts, or other small objects that may be accidentally dropped into the hopper. Metal contaminants can seriously damage the screw.

2.3.2 Barrel

The main components of a barrel are the barrel sheath, the screw, the thrust bearing, and a series of heater bands, which are used to melt the material. The purpose of the barrel is to house the screw and provide the delivery path to the die that forms the desired profile.

2.3.3 Extruder Screw

The screw is designed so that when it rotates, the resin pellets are metered forward by the flights, gradually melting and building up pressure along the way. Typical clearance between the flights of the screw and the barrel wall ranges from 0.003″ to 0.010″ (0.076 to 0.254 mm), depending on the size of the extruder. The depth of the flights, which is the distance from the outer edge of the flight to the shaft of the screw, varies, depending upon the section of the screw. The screw is divided into three sections or zones: the feed zone, transition zone, and the metering zone as shown in Figure 2.5.

In the feed zone, the screw has the largest flight depth so the unmelted plastic pellets can enter the barrel via the feed throat and be moved forward to the next zone. As the pellets enter the transition zone, the depth of the flights gradually decreases, which in turn increases the shear and pressure of the resin against the screw flights. The increased shear and pressure melts the pellets, which helps to reduce any imperfections in the feed, eliminate entrapped air pockets, and ensure a homogeneous resin melt. Finally, the resin enters the metering zone, where the flight depth remains the same as the smallest flight depth of the transition zone. The shear heat and pressure continue to build until the material reaches the die. Since the screw is an integral component of the extruder and the process, it is explained in further

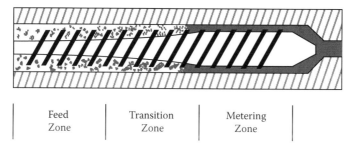

| Feed Zone | Transition Zone | Metering Zone |

FIGURE 2.5
Continuous feed screw.

detail. If you are only looking for basic information on the screw you can skip the following section.

2.3.3.1 An In-Depth Look at the Screw

A significant amount of research has been conducted pertaining to extruder screws and their design. The screw plays such a crucial role that the profile size, machine capacity, and part costs are determined by the extruder screw.

Extruder screws come in two basic configurations: either a single screw or twin screw option as shown in Figure 2.6. There are many types of individual screw designs, but they can all be categorized into these two main groups. A single screw extruder is typically right-handed and rotates counterclockwise, while a twin screw can rotate in either direction or in the same direction. Twin screws are utilized in high throughput applications such as master-batch compounding and large solid profiles like decking.

The type of screw selected depends on a number of variables, for example, material selection, profile size, and production volume. Screw factors to consider when selecting a screw include

Compression ratio—This is the ratio of the flight depth of the feed zone to the flight depth of the metering zone and is written as a standard ratio, for example, 2:1. As the compression ratio increases, so do the shear, heat, and potential for molded-in stress.

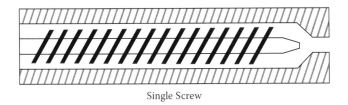

Single Screw

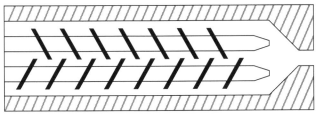

Twin Screw

FIGURE 2.6
Single and twin screw configurations.

Flight depth (also referred to as the channel depth)—This is a measurement of the distance between the outer edge of the flight and the shaft of the screw. When referring to flight depth, the output of the system and the shear are inversely proportional. As the flight depth increases, the output of the system increases, but the shear is decreased. This depth varies depending upon the zone of the screw.

Flight width—This is a measurement of the width of the individual screw flights. The typical width is 0.100″ (2.54 mm).

Length/diameter (L/D) ratio—Ratio of the flighted length of the screw divided by the outer diameter of the screw shaft. A common L/D ratio is 24:1, but this ratio can be as small as 15:1 or as large as 40:1.

Pitch—Distance from one screw flight to the next screw flight.

Helix angle—Angle of the screw flight measured by taking the angle between the plane perpendicular to the axis of the screw and the screw flight.

Screw profile—Measurement of the length of each zone of the screw. For example, increasing the feed zone length can increase the system output. Increasing the transition zone reduces the shear heat and increases the compression of the resin. Increasing the metering zone allows more pressure to build up before the die. If the length of the zone is decreased, the opposite will occur.[5]

2.3.4 Thrust Bearing

The thrust bearing connects the screw and motor linkage together and absorbs the force from the screw as it rotates against the plastic. It prevents the screw from moving backward in the barrel and absorbs the force generated by the screw as it rotates to melt the material. In high output applications, the pressure applied to the thrust bearing as a result of the higher speed of the rotating screw will cause it to wear faster than if it were run at lower speeds.

2.3.5 Heat Regions

The heat regions, also known as heater bands, maintain a constant temperature of the material in the barrel within a zone. They are not the primary sources of heat generation; the majority of the heat is shear heat created from the friction generated by the compression of the plastic in the barrel by the screw. In most cases, extruders have three or more independently controlled heater band regions to help maintain the desired temperature of the material extruded.

2.3.6 Breaker Plate and Screen Pack

The screen pack is a series of wire screens with varying mesh sizes used to filter out possible contaminants or unmelted resins before they reach the die and cause possible damage. The breaker plate is used to secure the screen pack in place. Additional strength is provided by the breaker plate, which is required because of the build-up in pressure. In some cases the pressure can be as high as 10,000 psi (69 MPa). The screen pack and breaker plate also provide back pressure to the barrel. Back pressure is necessary in the barrel to ensure a homogeneous melt of the resin.[6] As material is run through the extruder, the screen pack will begin to clog and the result will be an increase in the back pressure in the barrel. The screen pack will need to be replaced.

2.3.7 Die

A die is the shaper that is located at the output end of the extruder barrel. It forces the heated material to take on a specific shape as the plastic passes through it. Two types of dies are common in the extrusion industry. The first is a standard flat plate die, which is shown in Figure 2.7. The profile is simply cut into the plate. Flat plate dies are low in cost and can be fabricated quickly. The second is a streamline die, shown in Figure 2.8, which is more complicated than a plate die and is typically used for corrosive materials like PVC. Wire EDM is required to cut the die as it gradually tapers down to form the final profile. Extrusion dies are also heated above ambient temperature during processing to maintain a consistent temperature throughout the melted material. In addition, the heating minimizes the temperature effects, which can cause unpredictable dimensional variations in the profile.

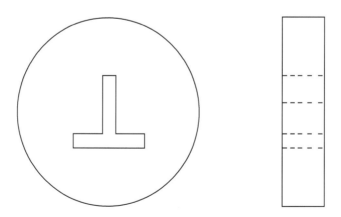

FIGURE 2.7
Flat plate die.

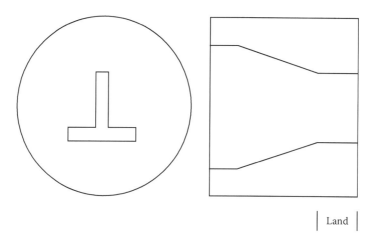

Land

FIGURE 2.8
Streamline die.

2.3.8 Cooling Zone

Typically in the cooling zone, the profile passes through a circulating water bath or a series of compressed air jets. Depending on the tolerance requirements of the profile, a series of reshaping plates may be located in the cooling zone. They are used to aid in controlling the shape and subsequently the final dimensions of the profile during the cooling process.

2.3.9 Reshaping Plates, Calibrator Plates, Vacuum Sizers

Since the profile actually starts to cool once it has left the die, calibrator plates and vacuum sizing plates may be required to provide additional support as the profile cools. The reshaping plates help maintain the desired final dimensions of the profile. Calibrator plates are a series of plates that the profile passes through as it cools. Figure 2.9 shows a set of calibrator plates. The first plate is oversized and each plate that follows is incrementally smaller until the profile is reduced to its final dimensions. If tighter tolerances are required they can be achieved either by adding additional reshaping plates or by reducing the spacing between the plates, or a combination of the two methods. Vacuum sizers, which are more expensive than calibrator plates, function in a similar way and are used on hollow profiles. The outer walls are drawn out using a vacuum.

2.3.10 Puller

Once the profile has cooled enough, it enters a pulling station. The puller is used to keep the profile moving through the entire process at a constant rate.

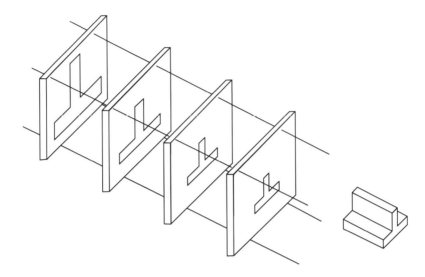

FIGURE 2.9
Calibrator plates.

2.3.11 Marking

Some profiles require identification once they have been cut to final length for traceability. Part marking such as a date code or lot number can be added easily and cost-effectively using a roller system, ink jet printing, or laser marker. Also during this stage, profile decorations such as wraps can be applied to the extrusion. This process applies bright colored finishes to the outside of the part.

2.3.12 Cutter

By the time the profile reaches the in-line saw it has been cooled enough to cut. There are two types of cuts: a low tolerance cut and a high tolerance cut. In a low tolerance cut, the profile is sawed with a tolerance of greater than ± 0.125″ (3.18 mm) of the overall length. In a high tolerance cut, the profile is cut to a longer manageable size and then transferred to a re-saw station where another cut is completed. It should be noted that a re-saw operation will increase the piece part price of the extrusion substantially, but tolerances of ± 0.06″ (1.5 mm) or better can be achieved. Tolerances also depend on the length of the part, the material processed, and the speed at which the material is run.

2.4 Tooling

2.4.1 Die Materials

Die wear is a factor in production. Dies are typically fabricated from hardened steel. If corrosive materials such as PVC are going to be processed, stainless steel can be used for tooling. For low volume, low tolerance parts, P20 steel and aluminum are used to fabricate tooling. Protective coatings are available to prolong tooling life when abrasive materials are extruded.

2.4.2 Die Fabrication

Extrusion dies are typically outsourced to professional toolmakers for fabrication versus being built in house. A small number of extruders possess the ability to fabricate tooling in house. Factors that can play into this decision are customer lead time requirements, simplicity of the dies compared to an injection mold (meaning no moving parts), lower dimensional accuracy, and the ability of the extruder to make modifications to the die if necessary.

Care should be taken while designing and fabricating an extrusion die. The cross sectional view of the die was previously shown in Figures 2.7 and 2.8. Particular attention must be paid to the land area (thickness) of the die because this area experiences the most pressure drop throughout the entire extruder. If the land length is too long, excessive back pressure could build up, reducing the output of the extruder and causing increased wear to the thrust bearing. If the land area is too short, the material will flow inconsistently and the profile dimensions will be harder to control.

Once the material is selected for the die, it is typically fabricated using a method called wire electrical discharge machining, better known as wire EDM. In some cases traditional machine methods can be used for simple die sets. Ideal parts for extrusion have uniform wall thickness in the cross or profile direction. The die is attached to the extruder breaker plate and the material is forced in a linear direction through the die at a relatively low pressure, under 100 psi, (0.69 MPa) as compared to injection molding pressures of 20,000 psi to 35,000 psi (138 MPa) to (241 MPa).

Preventive maintenance care is necessary to prolong tooling life as well as the extruder itself. In almost all cases, this is the responsibility of the supplier.

2.5 Materials

Extruders process resin in pellet form, as do most other plastic conversion processes. A variety of thermoplastics and thermoset materials are available for processing, for example, polyvinyl chloride (PVC) is widely used in the construction industry for window frames, siding, trim, and plumbing pipes; acrylonitrile butadiene styrene (ABS) is used for plumbing waste pipes; and polyethylene (PE) is used for insulating copper wire, drinking water tubing, and other flexible tubing (Table 2.1). Some materials will require drying prior to processing. If the material is not dried, air bubbles may be present on the surface of the part. Extrusion-grade resin characteristics feature a lower melt flow index viscosity, which relates to stiffness. The material must be rigid enough to support its own weight as it exits the die and cools to its final form.

There are a number of additives and fillers that are available for extrusion-grade resins. Some of the more common additives are wood flour, glass fibers, flame retardants, and ultraviolet stabilizers. Regardless of the conversion process, the addition of a filler or additive will have an impact on the resin process settings.

An interesting material to extrude is nylon, because it is a low viscosity material (it flows like water) and it will not support itself or maintain the part form as it exits the die. Another disadvantage of extruding nylon is that it requires drying prior to processing. An advantage of nylon is its high melting temperature and excellent chemical resistance characteristics.

TABLE 2.1

Common Extrusion Materials

Material	Properties	Sensitivity to Moisture	Applications
ABS	High impact Weather resistant Rigid	Yes	Plumbing Impact applications
HIPS/PS	Low process temperature High impact	No	Impact applications Low cost applications
PE—HDPE and LDPE	Chemical resistant Good impact High shrinkage rate	No	Drinking water tubing Low cost applications Medical
PP	Good impact High process temperature Highly chemical resistant	No	Chemical applications Tubing
PVC	Good impact Highly chemical resistant Rigid	No	Window industries Plumbing Chemical applications

2.6 Profile Extrusion Part Design Guidelines

When designing a part that utilizes the extrusion process it helps to keep in mind the following simple guidelines. Using these guidelines you should be able to design a part that most extruders can successfully quote without having to alter your design too much. Remember to check with the individual profile, coating, or sheet suppliers for their exact design guidelines and requirements. It should be noted that there is no industry standard for extrusion tolerances. This is done on a part-by-part basis. When processing regrind materials, it is necessary to open up the tolerances on the parts because they are not held in a mold until cooled as with other processes.

Wall sections (where t is equal to the nominal wall thickness)

- Uniform wall thickness (Figure 2.10) in the cross-sectional view is highly desired for dimensional accuracy. Extrusions designed in this manner display a consistent cooling pattern, and in turn provide more uniform part shrinkage and dimensional accuracy. These parts also demonstrate less of a tendency to curve or bow in a particular direction, and it is easier for the tooling maker to balance the flow of material through the die.
- Wall section thicknesses within 0.06″ (1.5 mm) of each other.
- No draft is required for extruded parts. This is a continuous process that is forced through a die.
- Part complexity is limited. Profile extrusion features need to be in the linear direction of flow. For example, undercuts cannot be placed in the cross-sectional plane of the part because the die has no way of forming the undercut while extruding at a constant rate. In the liner direction of flow, an undercut, illustrated in Figure 2.11, would run the entire length of the part.

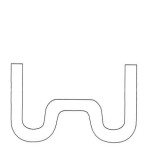
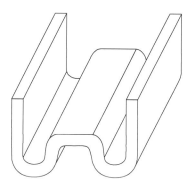

FIGURE 2.10
Design guide—uniform wall section.

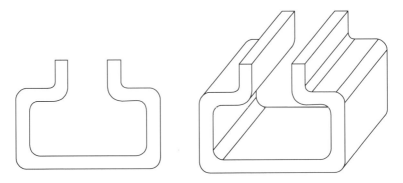

FIGURE 2.11
Design guide—undercut.

- Cost-effective at high volumes. Extrusion is an effective conversion process at high volumes because it is a continuous process. The higher the volume or linear feet the better. Once the equipment is set up, minimal expenses are required other than material and electricity to maintain large production runs.

- Typically cross-sectional part size is less than 12″ (30.48 cm), but larger cross sections are obtainable using a twin-screw extruder. Examples of small extrusions would be medical intravenous tubing, vinyl coated wires, ball-point pen ink tubes, and fiber optic cables. Larger extrusions include drainage pipes and sheet stock, which typically require a twin-screw extruder to deliver the high volume of material to the die.

- Tolerances on extruded parts are typically greater or looser than other plastic conversion processes because the part is not contained within a rigid form, like a mold, as the part cools. To assist in extruding parts with tight tolerances, the profile can be passed through a series of reshaping plates or forming stations as it cools.

Texture

- Parts are not textured.

Radii

- Minimum radius of 0.5 t of the nominal wall thickness with a 1 t preferred.

- No sharp edges.

Holes

- Secondary operations are required if holes or slots are designed into the part. When an extruded part requires a hole or slot in the linear direction, it is added offline as a secondary operation.

In most cases, the holes or slots are added using a punch press. In low volumes, the holes and slots are drilled or milled into the parts due to the costs associated with creating a shaped punch.

Part attachment methods

- Press fit/interference fit
- Ultrasonic welding
- Spin welding
- Adhesive
- Fasteners and inserts

2.7 How to Identify an Extruded Part

An extruded part can be identified by a number of key features. For example, in the cross-sectional view, look for a uniform wall thickness. Although parts can be extruded with varying wall sections, the difference between the largest and smallest wall section is typically within 0.06″ (1.5 mm). While looking at the cross section, can you see through to the other end, assuming that no additional parts like an end cap have been added. In the linear direction (direction of material flow), extrusions have the same wall thickness over the length of the part. The presence of streaks or die lines in the linear direction is also a major indicator of an extruded part. Before secondary operations, the cross section of the part will be the same at any point of the extrusion. The length of the part could also be an indicator of an extruded part. Part length over 8 ft (2.4 m), such as flexible tubing, rigid pipes, coated wires, or sheets is a good indicator that the part is an extrusion. The last major feature that is easily identifiable would be the lack of a gate or gate vestige. A gate would be present in an injection molded or blow molded part. Extruded parts are usually simple in design and lack a lot of part geometry.

- Wall section thicknesses within 0.06″ (1.5 mm) of each other
- Uniform profile
- Streaks or die lines in the linear direction
- Lack of a gate
- Parts of indefinite length
- Saw marks present on the cross-sectional view

2.8 Case Studies

2.8.1 Case Study 1—Similar Materials

To protect competitive advantage, the suppliers may rename materials once they are received in house. In this example, the request for quotes was returned with different specified materials, even though the initial request specified Georgia Gulf 5055 Rigid PVC-Black—the current material used.

To address this situation it was requested that the supplier quote the extruded parts using the current material, Georgia Gulf 5055 Rigid PVC-Black, and the alternative equivalent material that the supplier processes in high volume for other customers. There are two reasons for this request. First, the supplier should be able to provide its high volume material cheaper than the Georgia Gulf material because the supplier may purchase it by rail car or semi load versus by the gaylord. Second, the supplier should be comfortable processing a material run in high volume, as opposed to processing a material it may have never used before. Also remember to have the supplier submit the material data sheet with the quote. The material data sheets were used to compare the supplier-selected materials to the material currently processed.

The next step was to use the common material and compare other aspects of the quote such as tooling costs, cutting costs, part marking costs, storage and handling, minimum order quantities, and inspection intervals, etc., to each other. Once all aspects of the quotes were compared, the most cost-effective supplier, based on the common material, was selected. The next step was to compare the difference between the common material selected and the one the supplier had selected. Using simple math, the difference was either added or subtracted from the piece part price, depending on whether it was more or less than the selected material. It should be noted that in all cases the supplier-selected material was cheaper than the common material because it was purchased in higher volumes. Had this not been the case, the supplier would not have quoted its material in the first place! In the end, a supplier was selected along with the material and the parts have been in production since 2003.

2.8.2 Case Study 2—Pipes and Tubes

With respect to pipes and tubes, can a profile extruder produce them cost-effectively? While it is true profile extruders can produce both pipes and tubes, they tend not to do so because this is left to specialty suppliers. These businesses are cost-effective and efficient at extruding pipes and tubes. There are only two features to a pipe or tube extrusion die set: the outside of the profile and the inside which forms the hollow of the pipe/tube. As the die set extrudes a profile, both experience wear over time. To maintain the same dimensions, the extruder is required to fabricate a new die set.

For example, if a customer required a 1.00″ (2.54 cm) OD (outside diameter) pipe with a 0.06″ (0.15 cm) wall section and a ± 0.010″ (0.03 cm) tolerance, the die set would be a 1.00″ (2.54 cm) OD and a 0.875″ (2.22 cm) ID (inside diameter), as shown in Table 2.2.

As the die set wears and exceeds the 0.010″ (0.03 cm) tolerance, a new die set is fabricated to continue to meet the customer's requirements for a 1.00″ (2.54 cm) OD/0.875″ (2.22 cm) ID pipe. Instead of throwing out the original die set, it is machined to the next standard pipe sizes, in this case 1.125″ (2.86 cm) OD and 0.750″ (1.91 cm) ID. Now the extruder has expanded the number of pipe sizes that can be produced from one size pipe to four different size pipes; the original size 1.00″ (2.54 cm) OD/0.875″ (2.22 cm) ID, and three new sizes. Table 2.3 shows the addition of Die Set #002.

As Die Set #002 exceeds the acceptable tolerance limits it is machined to the next standard pipe size and the revision level is changed to Rev B. In addition to this, Die Set #003 is fabricated to replace Die Set #002. Table 2.4 shows the addition of Die Set #003, for a total of four different pipe configurations that are available for extrusion.

Over time, as the die sets wear they are converted to the next standard pipe size. Herein lies the opportunity: extruders that specialize in pipes and tubes are at a distinct advantage when it comes to tooling.

TABLE 2.2

Initial Die Set

Die Set	Tooling Rev	Outside Diameter	Inside Diameter
Die Set #001	Rev A	1.000″ (2.54 cm)	0.875″ (2.22 cm)

Total Number of Configurations

One configuration – [1.000″ (2.54 cm) OD/0.875″ (2.22 cm) ID]

TABLE 2.3

Alternate Die Configurations

Die Set	Tooling Rev	Outside Diameter	Inside Diameter
Die Set #001	Rev B	1.125″ (2.86 cm)	0.750″ (1.91 cm)
Die Set #002	Rev A	1.000″ (2.54 cm)	0.875″ (2.22 cm)

Total Number of Configurations

Four configurations –
  [1.125″ (2.86 cm) OD/0.875″ (2.22 cm) ID]
  [1.125″ (2.86 cm) OD/0.750″ (1.91 cm) ID]
  [1.000″ (2.54 cm) OD/0.875″ (2.22 cm) ID]
  [1.000″ (2.54 cm) OD/0.750″ (1.91 cm) ID]

TABLE 2.4

Alternate Die Configurations

Die Set	Tooling Rev	Outside Diameter	Inside Diameter
Die Set #001	Rev B	1.125″ (2.86 cm)	0.750″ (1.91 cm)
Die Set #002	Rev B	1.125″ (2.86 cm)	0.750″ (1.91 cm)
Die Set #003	Rev A	1.000″ (2.54 cm)	0.875″ (2.22 cm)

Total Number of Configurations

Four configurations –
[1.125″ (2.86 cm) OD/0.875″ (2.22 cm) ID]
[1.125″ (2.86 cm) OD/0.750″ (1.91 cm) ID]
[1.000″ (2.54 cm) OD/0.875″ (2.22 cm) ID]
[1.000″ (2.54 cm) OD/0.750″ (1.91 cm) ID]

In conclusion, pipes and tubes are an example of where a specialty extruder may hold a distinct advantage over custom profile extruders in terms of cost. They tend to have a large selection of die sets on hand and have become extremely efficient and cost-effective at die fabrication and processing. In some cases, the costs associated with pipe or tubing die fabrication can be minimized or even eliminated. For example, the EDM programming for a specific size may have already been completed, material for the desired die set can be made out of a tolerance size die so new die material is not purchased, and the set-up times are minimized due to past experience.

References

1. Leonard Nass and Charles A. Heiberger, *Encyclopedia of PVC: Compounding Processes, Product Design, and Specifications* (New York: Marcel Dekker, Inc., 1992).
2. James White, "History of Twin Screw Extrusion," Feedscrews, http://www.feedscrews.com (accessed October 12, 2007).
3. Oxford Plastics, Inc., "A History of Polyethylene Pipe," http://www.oxford-plasticsinc.com (accessed October 22, 2007).
4. Bad Fads, "Hula Hoops," http://www.badfads.com (accessed October 23, 2007).
5. Joe Slenk, "Design Variables," Ferris, http://www.ferris.edu (accessed July 7, 2008).
6. Wikipedia, "Plastics Extrusion," http://en.wikipedia.org/wiki/Plastics_extrusion (accessed April 20, 2008).

3

Blow Molding

Blow Molding Process Key Characteristics

Volume	High
Material selection	Moderate
Part cost	Low
Part geometry	Simple
Part size	Small to very large
Tool cost	Medium to high
Cycle time	Seconds
Labor	Automatic

For a list of other conversion process characteristics, see Appendix B.

3.1 Process Overview

Blow molding is a low pressure process (25–350 psi) or (0.17–2.41 MPa) for forming hollow thermoplastic parts. An extruded parison or preform is surrounded by a mold and pressurized to take on the contours of the mold cavity. This process is similar to injection molding in that plastic resin is put into a hopper where it is fed through a heated barrel by a continuous screw. As the resin moves forward through the barrel, it melts and is then extruded through a vertical head die as a hollow tube called a parison. As the parison is extruded, the mold halves close around the parison and air is forced into the inside of the parison by means of a blow pin, which causes it to expand and take on the contours of the mold. The part is allowed to cool and is then ejected from the mold once it opens. Excess material is typically trimmed from the part by pinch-off surfaces in each mold half. In some cases, excess material is trimmed as a secondary operation. The cycle is then repeated. Figure 3.1A depicts the blow molding process. Photographs of blow machines are shown in Figures 3.1B and 3.1C.

Why would you use this process? It is well suited for the high volume manufacturing of small to medium sized hollow parts, like bottles.

The most familiar applications of this process include

- Automotive
- Electronics

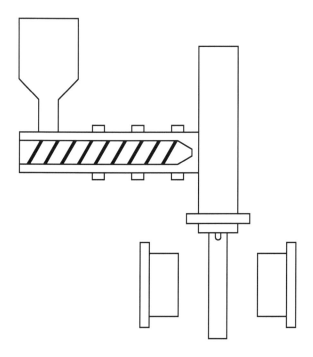

FIGURE 3.1A
Diagram of a blow molding machine.

- Marine
- Medical
- Sporting goods
- Toys
- Bottles
- Storage drums

3.1.1 Variations of the Blow Molding Process

As demand for blow molded products increased, capacity became an issue. Additional hybrid processes were developed to improve capacity and provide expanded design capability. There are three primary methods: extrusion blow molding (EBM), injection blow molding (IBM), and stretch blow molding (SBM).

Extrusion blow molding was the initial process developed to create small hollow products. As described before, the plastic material is extruded in the form of a hollow tube or parison. In most cases, this is a continuous process but in some machines the parison is extruded intermittently. Using two molds in a shuttle motion, as illustrated in Figure 3.2, one mold closes around the parison, pinching off excess material. It is shuttled to a second position

FIGURE 3.1B
Bekum blow molding machine. *Source:* Photo courtesy of MGS Mfg. Group.

FIGURE 3.1C
Davis-Standard blow molding machine. *Source:* Photo courtesy of MGS Mfg. Group.

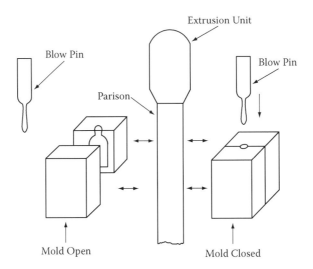

FIGURE 3.2
Diagram of the shuttle molding process.

where the part is inflated and cooled. The mold opens for part removal and moves back to the position underneath the parison that is being extruded to receive material for the next cycle. The molds alternate between the parison and blowing stations continuously. Figure 3.3 shows the extrusion blow molding process. Intermittent extruded blow molding features an accumulator or reservoir where the melted resin accumulates. When enough material has been melted, a hydraulic ram activates, forcing the material through the die where the mold will clamp around the parison.

The second blow molding process variation is called injection blow molding and is illustrated in Figure 3.4. This method is extensively used for capped bottles. It starts with the injection molding of a preform onto a core pin. The preform includes the neck with external threads and a thick tube attached to

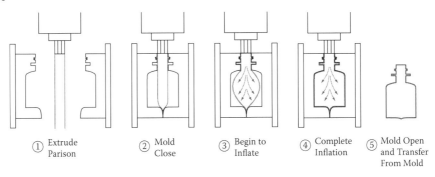

FIGURE 3.3
Variation—extrusion blow molding.

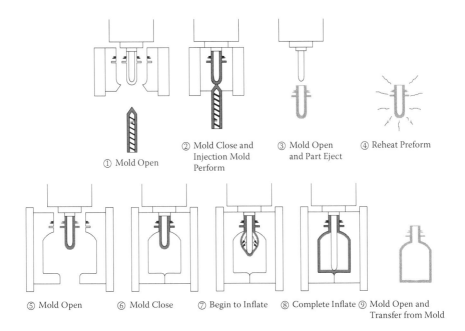

FIGURE 3.4
Variation—injection blow molding.

the neck. The tube contains enough material to form the body of the bottle in the blow stage of this process. Next, the mold opens and the preform is transferred to the blowing station. This can be done while the preform is still hot or it can be reheated. If it is reheated, it will be heated above the material's glass transition temperature. A second mold, with contours reflecting the final part, closes around the preform. It is then inflated and allowed to cool before the mold opens and the part is ejected.

Stretch blow molding is the third variation of blow molding and a diagram is shown in Figure 3.5. Like injection blow molding, stretch blow molding can utilize a preform component. In this process, the tube preform is placed into a press equipped with a moveable mandrel. The mold closes around the preform heating the tube portion to the glass transition temperature. Simultaneously, the mandrel advances to stretch the material while it is inflated. A visible mark from the mandrel can be seen at the bottom of the part. After the part has cooled, it is ejected and the cycle is repeated. This process is commonly used in the manufacturing of soda bottles. Instead of shipping large empty bottles, some bottlers do the blow molding of the preform in house and then fill the bottles immediately afterward.

Two recent advancements in blow molding technology include programmable parison profiles and multilayer blow molding. Newer blow molding machines allow the operator to customize the profile of the parison. For example, if the bottle is tapered, the parison can be programmed to be

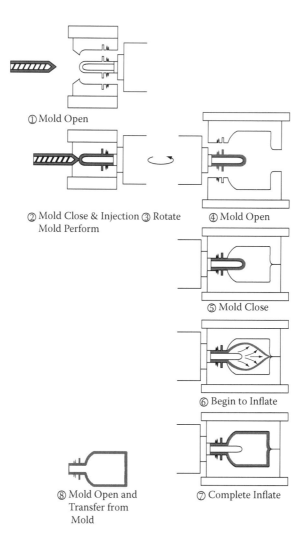

①Mold Open

②Mold Close & Injection ③Rotate
Mold Perform ④Mold Open

⑤Mold Close

⑥Begin to Inflate

⑧Mold Open and ⑦Complete Inflate
Transfer from
Mold

FIGURE 3.5
Variation—stretch blow molding.

thicker at the wider dimension. This helps to maintain a constant wall thickness throughout the entire part once it has been inflated.

Multilayer blow molding equipment features additional extruder units and accumulators, which are shown in Figure 3.6. This is a modification of the equipment used in the extrusion process, where multiple materials are co-extruded in layers. Multilayer bottles offer a variety of colors and property enhancements such as light blocking or additional permeation or barrier layer protection.

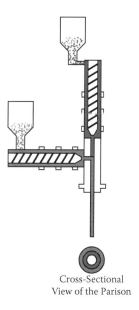

Cross-Sectional
View of the Parison

FIGURE 3.6
Multilayer molding.

3.2 A Brief History of Blow Molding

The origins of blow molding can be traced back to ancient times when molten glass was blown into various shapes, which became bottles. Modern-day plastic blow molding started in the 1930s when equipment was developed based on the principles of glass blowing. The material used was cellulose nitrate and later cellulose acetate. The problem with these bottles was that they were not much of an improvement over the traditional glass bottles. It was not until the introduction of low density polyethylene (LDPE) in the 1940s that plastic bottles began to replace traditional glass bottles. By the 1950s blow molding grades of high density polyethylene (HDPE) and polypropylene (PP) became commercially available. These materials would forever replace the current storage medium of glass or metal in some applications and help increase the demand for blow molded goods. Low material cost, good moisture barrier, and chemical resistance were impressive characteristics of these materials; however, there were issues with carbonated liquids and solvents. The solution was discovered in the 1970s, with the introduction of polyethylene terephthalate (PETE), which exhibited superior barrier layer properties. This material went on to revolutionize the carbonated beverage industry. Soda bottles now come in a wide variety of shapes, sizes, and colors, complete with contoured gripping features and colorful advertising.

In the late 1990s the beverage and soft drink industry experienced another trend change when the bottled water craze spread across America and other parts of the world, taking demand from zero units in 1977 to over 10 billion by 1999.[1] By 2005, this number had grown to 26 billion bottles of water consumed per year.[2] Now, this phenomenon has become a major recycling issue as the majority of these bottles end up in landfills.

3.3 Equipment

Typically, there are six steps involved in blow molding thermoplastics: heating the resin in the extruder, forming the parison in the parison die, blowing or forming of the part in the mold, cooling the molded part, removing the part from the mold, and trimming the excess material away from the part. Additional steps such as labeling and filling can be completed in-line or moved to another station. Figure 3.7 illustrates a basic blow molding equipment set-up. Detailed descriptions of the main components of the machine are listed below.

1 – Hopper
2 – Barrel
3 – Screw
4 – Heat regions

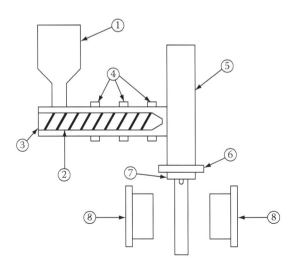

FIGURE 3.7
Blow molding equipment diagram.

5 – Accumulator

6 – Breaker plates and screen pack

7 – Parison die

8 – Mold

3.3.1 Hopper

The hopper, which is also used on injection molding and extrusion equipment, funnels unmelted plastic pellets, by gravity, to the feed section of the barrel. Some hoppers will have a transparent window to view the material level. Material can be added manually or with an attached vacuum system for high throughput applications. Hoppers are covered to prevent possible contamination and also feature a magnetic screen above the entrance to the throat of the barrel to catch any metal fines, chips, bolts, or other small objects that may be accidentally dropped into the hopper. Metal contaminants can seriously damage the screw.

3.3.2 Barrel

The main components of a barrel are the barrel sheath, the screw, the thrust bearing, and a series of heater bands, which are used to melt the material. The purpose of the barrel is to house the screw and provide the delivery path to the parison die.

3.3.3 Continuous Feed Screw

The screw is designed so that when it rotates, the resin pellets are metered forward by the flights, gradually melting and building up pressure along the way. Typical clearance between the flights of the screw and the barrel wall ranges from 0.003″ to 0.010″ (0.076 to 0.254 mm), depending on the size of the extruder. The depth of the flights, which is the distance from the outer edge of the flight to the shaft of the screw, varies based on the section of the screw. The screw is divided into three sections or zones: feed zone, transition zone, and metering zone, as shown in Figure 3.8.

In the feed zone, the screw has the largest flight depth so the unmelted plastic pellets can enter the barrel via the feed throat and be moved forward to the next zone. As the pellets enter the transition zone, the depth of the flights gradually decreases which in turn increases the shear and pressure of the resin against the screw flights. The increased shear and pressure melts the pellets, which helps to reduce any imperfections in the feed, eliminate entrapped air pockets, and ensure a homogeneous resin melt. Finally, the resin enters the metering zone, where the flight depth remains the same as the smallest flight depth of the transition zone. The shear heat and pressure continue to build until the material reaches the die. Since the screw is an

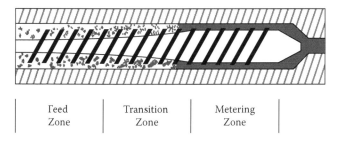

Feed	Transition	Metering
Zone	Zone	Zone

FIGURE 3.8
Continuous feed screw.

integral component of the extruder and the process, it is explained in further detail in Chapter 2 – Extrusion.

3.3.4 Heat Regions

The heat regions, controlled by heater bands, maintain a constant temperature of the material in the barrel within a zone. They are not the primary sources of heat generation; the majority of the heat is shear heat created from the friction generated by the compression of the plastic pellets in the barrel by the screw. In most cases, blow molding barrels have three or more independently controlled heater band regions to help maintain the desired temperature of the material extruded through the barrel.

3.3.5 Accumulator

Accumulators are not used with all blow molding machines, but are required for intermittent blow molding processes. Resin is melted and is accumulated until a hydraulic plunger is used to force the material through the die to create the parison.

3.3.6 Breaker Plates and Screen Pack

The screen pack is a series of wire screens with varying mesh sizes used to filter out possible contaminants or unmelted resins before they reach the die and cause possible damage. The breaker plate is used to secure the screen pack in place. Additional strength is provided by the breaker plate, which is required because of the build-up in pressure.

3.3.7 Parison Die

This plate, with the programmable mandrel, forms the shape and thickness of the extruded tube.

3.3.8 Mold

The mold is typically a two-piece device and if threads are used, this area may be an insert into the tooling. Molds can feature parison pinch-off points near the bottom to remove excess parison from the mold during the molding cycle.

3.4 Tooling

3.4.1 Mold Materials

Aluminum, beryllium-copper, steel, and stainless steel are used to fabricate blow mold tooling, with aluminum and beryllium-copper leading the way. These materials are selected primarily for their excellent heat transfer properties, but also their low cost and durability. In most applications, the capacity of a machine is limited by the ability of the equipment to cool the part so it can be ejected. Therefore, heat transfer becomes critical to this process. Aluminum tooling is typically used when processing HDPE, while beryllium-copper or stainless steel is used for PVC parts. Some processes use CO_2 gas to inflate the parison and increase cooling.

3.4.2 Mold Fabrication

Blow molding tooling is machined in symmetrical halves using conventional machining methods such as computer numerical control (CNC) machining and EDM. Care must be taken when fabricating molds from beryllium-copper, as the dust can cause health issues. Molds also feature cooling systems, ejection systems, and vents. Unique to blow molds is a feature called a pinch-off area, which serves to automatically separate the parison from the finished part.

Inserts, which are replaceable components of the mold, can be added in high wear areas. Inserts are also used in the thread and neck areas of the bottle. This allows for the same profile bottle to be molded with a different set of threads on the neck.

3.5 Materials

Amorphous materials are easier to process than crystalline materials because of their wide melt temperature range. Materials used in the blow molding process start in the same form as injection molding and extrusion—pellets. Blow molding grade materials, however, have lowered melt flow indexes. Materials with chemical resistance and impact are widely used.

PETE, PE (both HDPE and LDPE), and PVC are the most common materials used in the blow molding process.[3] Table 3.1 lists some of their key material properties and applications.

Other materials not listed can be processed, and they typically exhibit a lower melt flow rate (i.e., MFR = 2), which gives them higher melt strength. The melt strength is important because the parison must support its weight during extrusion and the material must resist tearing when the cavity is pressurized. While the melt strength does not appear on the resin supplier's material data sheets, the melt flow rate will.

In 1988, the Society of the Plastics Industry, introduced the seven-code resin identification system to help increase the awareness of recycling plastic.[4] To simplify container thread and cap designs, a standardized specification has been developed by the Society of the Plastics Industry and the Glass Packaging Institute, which specifies the outside diameter of the thread, the outside diameter of the neck, the inner diameter of the neck, the orientation of the closure, and the height of the finished neck. This specification allows bottlers to set up the filling equipment without having to guess at the type of bottle neck or threads. This also helps to standardize the caps for specific threads.

TABLE 3.1

Common Blow Molding Materials

Material	Properties	Sensitivity to Moisture	Applications
PETE	Excellent barrier properties Good clarity Excellent impact properties	No	Beverage bottles
PE—HDPE and LDPE	Chemical resistant Translucent Good impact High shrinkage rate	No	Chemical storage Food and liquid storage
PVC	Good impact Highly chemical resistant Rigid	No	Chemical applications

3.6 Blow Molding Part Design Guidelines

Design guidelines for blow molded parts are not as complicated as those for injection molding, and the plastics industry has standardized bottle neck sizes and threads in an attempt to minimize design problems. Consulting with your molder will help to ensure that the part can be molded to meet your specifications.

Wall sections (where t is equal to the nominal wall thickness)

- Wall sections will vary in blow molding depending on the geometry of the part. Parts cannot easily be uniformly maintained as they can be in injection molded parts.
- Typically no draft is required on part sides, only the top and bottom require a minimum 1° draft.
- Draft texture 1° per 0.001″ (0.025 mm) of texture depth.
- Texture can be present on the sides of the bottle.

Bottle neck

- Bottle neck specifications have been standardized by the Society of the Plastics Industry and the Glass Packaging Institute to make it easier and more cost effective to blow mold plastic bottles and mating closures. This standard also provides a specification for the minimum clearance for filling tubes.

Radii

- Minimum radius of 1 t.
- Chamfer preferred over radii if possible.
- No sharp edges.

Ribs

- Ribs may be used to strengthen parts.

Holes

- Holes are typically centered on one end of the part.
- Additional holes can be added as a secondary operation.

Corners

- Tend to be thinner than the nominal wall section.

Part attachment methods

- Snap fits.
- Ultrasonic welding.
- Spin welding.

- Adhesive.
- Fasteners and inserts can be installed using ultrasonic welding.

3.7 How to Identify a Blow Molded Part

Below is a list of some key features of blow molded parts.

- Hollow construction
- Parting line down the center of the part
- Absence of ejector pin marks
- Variable wall thickness at the corners

3.8 Case Studies

3.8.1 Case Study 1—Gasoline Containers

The first commercially available gasoline cans were fabricated out of sheet metal. The cans could store 1 to 5 gallons of gasoline and featured two threaded openings; one was used as the pour spout and the other as a vent for the can. The cans also had attached carrying handles. These cans did have some drawbacks, however. For instance, the caps, if dented, were extremely difficult to remove or secure in place and they could get lost since they were not attached to the can. Also, the cans could get dented or would rust. Furthermore, the cans demonstrated sealing issues during pouring gasoline that would result in spills.

Once the press size and control of the presses improved, plastic gasoline containers were developed in the 1980s and quickly began to replace metal gasoline cans. Today, only plastic gasoline containers are available for storage of gasoline in 1 to 5 gallon amounts. Plastic gasoline containers have an integrated handle, a threaded cap, which holds the pouring spout, and a vented cap with a retaining ring to secure the cap to the can (Figure 3.9). Plastic containers have two drawbacks. The first is that these containers can build up a static charge, which can cause the vapors inside the container to ignite. The second drawback is that gas vapor expands in hot weather and causes the containers to swell. Overall, plastic containers are more user friendly with easier to hold handles and convenient cap tethers, but some people will always prefer metal cans.

FIGURE 3.9
Metal and plastic gasoline containers.

3.8.2 Case Study 2—Container with Annular Snap Fit Cover

Applications that require a container and a cover can be manufactured using the blow molding process. An annular snap fit feature can be designed into both the container and the cover. In this design, the container and cover are molded as one piece. The part is then moved to a re-saw station where the two pieces are separated and excess material is removed from the parts. Any burrs and sharp edges are then removed from the cut surface. Depending on the design, a number of containers can be stacked inside each other for shipping.

3.8.3 Case Study 3—Traffic Construction Barrels

Construction zones are required to be clearly marked in order to protect the job site workers and motorists. Large safety barrels were designed to satisfy this need. The barrels have provisions to allow additional weight. Recycled tires are used to form a ring that is placed over the barrel to stabilize it when in use (Figure 3.10).

FIGURE 3.10
Construction barrel with recycled tire ring at the base for extra weight.

References

1. Wikipedia, "Blow Molding," http://en.wikipedia.org/wiki/Blow_molding (accessed June 6, 2008).
2. Pat Franklin, "Down the Drain," Container Recycling Institute, http://www.container-recycling.org (accessed March 9, 2008).
3. Earth Odyssey, "Symbols," http://www.earthodyssey.com (accessed March 12, 2008).
4. IDES, "Resin Identification Codes—Plastic Recycling Codes," http://www.ides.com (accessed March 14, 2008).

4

Thermoforming

Thermoforming Process Key Characteristics

	Sheet Stock	**Roll Stock**
Volume	Low	High
Material selection	Moderate	Moderate
Part cost	Medium	Very low
Part geometry	Simple	Simple
Part size	Large to very large	Small to medium
Tool cost	Low to high	Low to high
Cycle time	Seconds to minutes	Seconds
Labor	Manual	Automatic

For a list of other conversion process characteristics, see Appendix B.

4.1 Process Overview

Thermoforming is a low pressure process of converting single plastic sheets or continuous roll stock. In this process, materials are classified into two categories based on thickness. Materials thicker than 0.010″ (0.254 mm) are referred to as sheets while materials thinner than 0.010″ (0.254 mm) are called films. This chapter focuses on sheet processing since it is the more common of the two material categories. The material is clamped into a rigid frame and moved to the heating stage where it is heated until it becomes semi-pliable. It is then drawn onto a mold form using one of the following processes: vacuum pressure, forced air (pressure), or mechanical assisted plugs. The part cools very quickly and excess material is trimmed. Finally, the formed part is stacked or, in food packaging applications, the parts may be sent through an auto-fill station. Figure 4.1 shows continuous feed and individual sheet systems.

This process differs from the other plastic conversion processes in that the material is not completely melted during the forming cycle; it is only heated to its glass transition temperature (T_g) where it becomes semi-pliable. This also allows for extremely short forming times for thin films. Approximately 75% of the thermoforming equipment is dedicated to the packaging industry.[1]

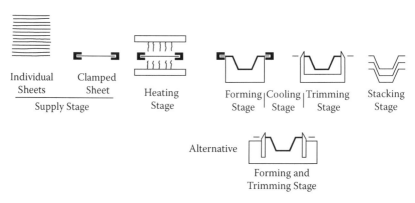

FIGURE 4.1

Diagram of the individual sheet and continuous roll thermoforming process.

However, thermoformed parts come in all shapes and sizes and can be seen elsewhere.

Why would you use this process? It is a very versatile process that is well suited for simple part geometries and anywhere from low to high part volumes. Compared to other conversion processes, thermoforming has one of the largest ranges in part sizes that can be formed, 1″ × 1″ (2.54 × 2.54 cm) or parts that are larger than 8′ × 8′ (2.4 × 2.4 m). Also, parts that will be molded using another plastic conversion process can be prototyped using the thermoforming process.

The most familiar applications of this process include

- Food and beverage containers
- Packaging
- Electronics
- Automotive
- Medical
- Agriculture
- Toys

4.1.1 Variations of the Thermoforming Process

There are four key characteristics that identify the thermoforming process. The first is the form of the material stock used in the process. Unlike the small pellets used in injection molding, blow molding, and extrusion, individual sheets or continuous rolls are supplied to the equipment.

The second characteristic unique to thermoforming is the type of mold that forms the part. High pressure processes, such as injection molding require two mold halves and a clamping system. Thermoformed parts are drawn into or onto one mold half. These molds are referred to as male and female. The male or positive molds have projecting surfaces, while the female or negative molds, have concave surfaces. Figure 4.2 shows the cross-sectional views of both types of molds. Different product features determine the choice of mold

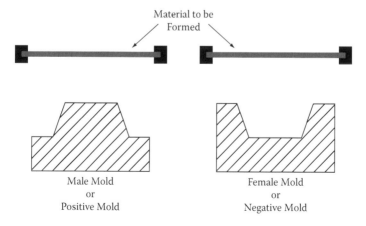

FIGURE 4.2
Example of a male (positive) mold and a female (negative) mold.

style. For example, a company logo or texture would require a female mold, while a recycling symbol may appear inside a part requiring a male mold.

The third identifying characteristic relates to the process itself. Figure 4.3 shows how the mold is positioned in relation to the material prior to the sheet formation. Specifically, the process is considered forming up when the material is positioned below the mold and it is drawn up to the mold. Forming down is the opposite of forming up; the material is positioned above the mold and drawn down to the mold. The sheet is heated until it sags, which pre-stretches it before forming. The sag of the sheet can be useful if forming down into a female mold or forming up into a male mold.[2]

The fourth characteristic of thermoforming relates to the process used to draw the sheets onto the mold. Parts can be formed either by vacuum pressure, air pressure, or plug assist, as illustrated in Figure 4.4. If the part features have simple contours, do not require a deep draw, and use sheet stock typically less than 0.100″ (2.54 mm), vacuum forming is the preferred manufacturing method. In vacuum forming applications, a clamping mechanism

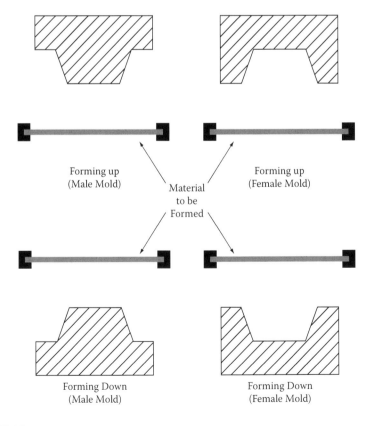

FIGURE 4.3
Forming up and forming down using male and female molds.

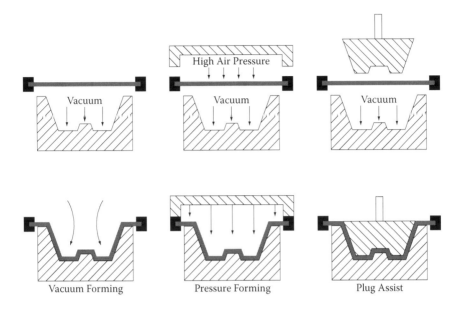

FIGURE 4.4
Forming methods—vacuum forming, pressure forming, and mechanical assist plug.

is used to hold the material as it moves through the thermoforming stations. After loading the material, it is moved to the heating elements. Once it has been heated, it is positioned over the mold and formed. Small holes, which are referred to as vents, ranging in size from 0.015" to 0.025" (0.381 to 0.635 mm) are drilled into the mold cavity to ensure that the trapped air is vented completely from the mold when the vacuum is applied. When the material comes in contact with the mold, the thickness of the material will remain constant at that particular location. However, material next to the fixed thickness must stretch to come in contact with the mold. This will result in thinner walls. Thinning typically occurs in corners and along the draw direction. Vacuum forming is limited to a maximum of 14.7 psi (1 atmosphere).

Pressure forming is a variation of vacuum forming where air pressure, as opposed to vacuum is utilized to form the part. The clamping mechanism secures the material as it is heated and sealed over the mold. Then compressed air is forced onto the side opposite of the mold, causing the material to form around the mold. This is the difference between pressure forming and vacuum forming. Air trapped between the material and the mold is allowed to escape through a series of vents located in the mold. Using this method of thermoforming tighter tolerance parts with greater part detail can be achieved with shorter cycle times than vacuum forming. Note that air pressure used in the process can greatly exceed the 14.7 psi (1 atmosphere) limits of vacuum forming.

In the plug assist method, material is handled in the same manner as the previous two methods with the exception of the plug. It is used in conjunction with female molds only where the plug assist is positioned over the material and looks like the mold except for the fact that is it 0.188" to 0.500" (4.77 to 12.70 mm) smaller, depending on the size of the part formed. This method is useful for forming deep draw parts. Plugs can be fabricated from similar materials as the mold, for example, aluminum, steel, plaster, etc. In some cases it may be necessary to heat the plug to prevent the material from cooling as it comes in contact with the plug, but before it has been completely formed.

4.2 A Brief History of Thermoforming

Thermoforming is one of the oldest and most common forms of plastic conversion processes. Primitive methods of thermoforming can be dated back to the Roman Empire when tortoise shells and hot oils were used to form food utensils. Consumer products such as baby rattles and teething rings have been made using the thermoforming process as early as 1890.[3] Thermoforming technology remained relatively unchanged until the 1930s, when thermoforming as well as other plastic conversion processes experienced a rapid expansion in uses and products as new materials became available.

In 1935, Otto Rohm, founder of the present day Rohm & Haas company, developed polymethyl methacrylate, better known as acrylic in the United States and PMMA in Europe and other countries. A year later, it was discovered that it could be thermoformed and it was used for the cockpit canopies of German airplanes. A great deal of research went into heavy gauge thermoforming of defect-free canopies. Not only were these new cockpit canopies more impact resistant than glass, they were also lighter in weight, and were more damage resistant when shipped around the world as repair parts. Later, the United States employed heavy gauge thermoforming technology on the B-29 bomber for cockpit canopies, the nose, and gun turrets. Contoured relief maps were also produced using this process.[4]

After World War II, thermoforming again experienced an increase in usage and applications. By the mid 1950s, thermoformed parts could be seen in the medical field as a replacement for traditional wood and leather prosthetics, as well as in blister packages and food storage containers. During the 1980s and 1990s food and beverage containers increased the demand for thermoforming, making it the process of choice. Today, thermoformed parts are all around us from storage cabinet sides and doors, luggage, cargo bins, retail packaging, food packaging, spas, shields, exterior window frames, privacy fences, machine guards, and more.

4.3 Equipment

Equipment used in the thermoforming process is listed below and shown in Figure 4.5. The equipment is grouped into stages of operation to simplify the explanation.

1 – Supply
2 – Heating
3 – Forming
4 – Cooling
5 – Cutting and trimming
6 – Stacking
7 – Waste take-up

4.3.1 Supply Stage

Plastic material is supplied in two standard forms, the individual sheet and the roll. Individual sheets are used in low volume applications or for very large parts. Continuous feed rolls are often used in high volume applications, and in some cases, the extruder is connected directly to the thermoforming equipment, providing a continuous supply of material. This allows the thermoformer a high degree of control over the sheet thickness fed into the equipment. The material is fed to a clamping mechanism that keeps it level as it enters the heating stage and at a constant distance from the heating elements during each cycle.

4.3.2 Heating Stage

Care must be exercised to uniformly heat the sheet. It is critical that the sheet is not heated above its maximum processing temperature because the material can degrade or worse, melt or burn. Conversely, if the material is formed under its minimum processing temperature the material can be overly stressed during the forming stage, which can result in stress

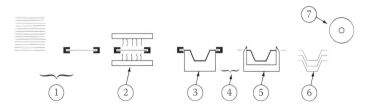

FIGURE 4.5
Thermoforming stages (with arrow callouts for each stage).

cracking and lower performance characteristics in the formed part. A variety of heating element assemblies (ceramic heaters, radiant panel heaters, etc.) are used in thermoforming equipment and they all perform differently to some extent. Zone temperature controls are used to create uniform thermal heating especially near the frame clamp, which tends to be a heat sink.

4.3.3 Forming Stage

Molds are used to create the forms of the desired parts. Unlike injection molding or blow molding processes, which require two halves of the mold to create a part, thermoforming may only require a single half of the mold to create the product. This helps to keep thermoforming tooling costs down. As described earlier in this chapter, parts are formed using one of three methods (depending on the required detail): vacuum, compressed air, or a mechanical assisted plug.

During the forming stage, the material comes into contact with the mold and the part is formed. Figure 4.6 illustrates various stages of the forming process in a forming "down" configuration for both male (positive) and female (negative) molds. Vents, which are not shown in the figure, are used to remove trapped air within the mold. Mold temperature has an impact on part shrinkage; as the mold temperature increases, so does the shrinkage of the part.

4.3.4 Cooling Stage

Ambient air, compressed air, CO_2, or water sprays can be used to cool parts. To maintain consistency from part to part, cooling lines can be run throughout the mold. Fans or compressed air are also used because they are usually readily available at the manufacturing site. These methods can be used because the material is only heated to its glass transition temperature (T_g) where it becomes semi-pliable versus completely melting as in other plastic conversion processes. When nonmetal molds are used, it can be difficult to control the cooling and they should only be used for short runs or when tight tolerances are not required.

4.3.5 Cutting/Trimming Stage

Trimming excess material can be completed either in-line or off-line. Tooling used to trim the excess material includes band saws, hand tool cutters such as knives, punch presses, shears, steel rule dies, routers, and water jets. Continuous feed operations trim formed parts in-line, while individual sheet feed operations have the flexibility to be trimmed both in-line and off-line. It is important to trim the parts at the same temperature every time to ensure that the final dimensions are maintained from part to part. Tolerances on the

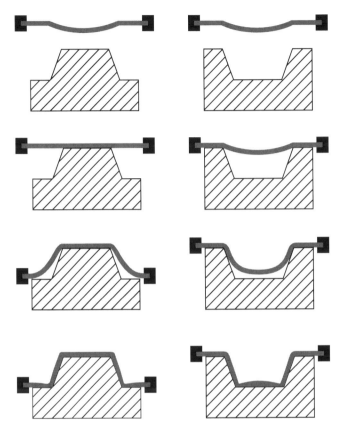

FIGURE 4.6
Comparison of male and female molds in various stages of forming.

trimming or cutting range from ±0.002″ (±0.051 mm) to ±0.01″ (±0.254 mm), depending on the part and can vary from thermoformer to thermoformer.

4.3.6 Stacking Stage

Once the part has cooled, one of two things can occur. The individual pieces can be stacked or the parts can be filled and sealed. A number of snack food products, such as pudding and yogurt, are manufactured using the fill and seal method.

4.3.7 Waste Take-Up Stage

In a continuous feed thermoforming process, the material remaining after the part has been removed can be recycled if it has been processed within the processing temperature window of the given material. Material processed outside the window should not be used as temperature can drastically affect the quality of the new material.

4.3.8 Post Forming

In addition to trimming, other post forming operations can occur.

- Priming and painting
- Pad printing designs or text
- Bonding posts for fasteners
- Drilling holes
- Routing openings
- Attachment of other parts
- Buffing and polishing
- Labeling
- Filling and sealing

4.4 Tooling

4.4.1 Mold Materials

A variety of materials can be used to fabricate a thermoform mold economically. Since this is a low pressure process, materials such as wood, plaster, epoxy, and rapid prototyping materials (stereolithography) can adequately function in low volume applications. When volumes increase, aluminum or steel should be utilized for tooling. Table 4.1 lists some of the more common materials used to fabricate thermoforming molds.

TABLE 4.1

Thermoform Mold Materials

Material	Estimated Tooling Life	Cost	Sensitivity to Moisture
Wood	Varies 1000–10,000	Low	High
Plaster	Under 100	Low	High
Epoxy	Varies	Low	N/A
SLA	Varies	Low	N/A
Aluminum	Over 1,000,000	High	Low
Steel	Over 1,000,000	High	Low

4.4.2 Mold Fabrication

Molds are machined from aluminum, steel, wood, and other materials. Figure 4.7 shows isometric views of a male mold and a female mold, both with vents. The tooling is built oversized to compensate for material shrinkage, which depends on the material processed. Vent holes are machined throughout the surface of the mold and in particular near the edges and internal corners where air can become trapped between the material and the mold. If cooling is required, cooling lines can be added to the mold.

Parts that require small undercuts can be formed using one-piece tooling. As the material cools in a female mold it shrinks away from the mold, allowing it to be easily removed from the tool. In some cases, undercuts can be stripped from a male mold, but larger sized undercuts can be achieved using female molds. Part designs that require very large undercuts can be formed utilizing multiple-piece tooling as shown in Figure 4.8. When multiple-piece tooling is used, parts within the mold move to allow the undercut to be removed from the mold. Some materials can be removed from the tooling easier than other materials.

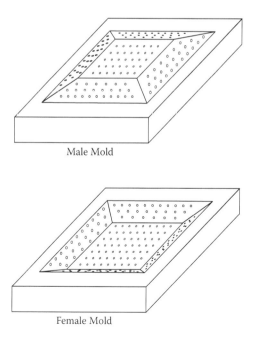

Male Mold

Female Mold

FIGURE 4.7
Isometric view of male and female thermoforming molds.

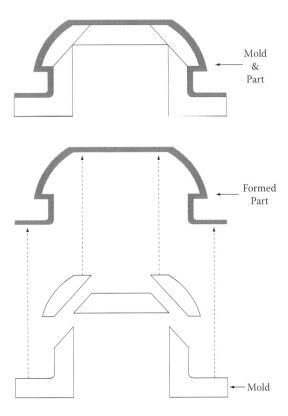

FIGURE 4.8
Multi-piece tooling for creating undercuts.

4.5 Materials

The materials used in this process are different from those used in other conversion processes in two ways. The first is that the raw materials come in sheets or rolls versus pellets, powders, or liquids. These materials have already been processed once by another conversion process, namely by extrusion to create the sheet, which is then used in the thermoforming process to create parts. The second difference is that the materials are not melted completely when processed. Most thermoplastic materials can be processed, but the preferred materials are amorphous (ABS, HIPS, PS, PVC, etc.), and do not have a sharp melting point like crystalline materials (PE, PP, PA, etc.). When amorphous materials are heated to their respective glass transition temperature (T_g) they become soft and pliable. Due to the random order of the molecular chains of amorphous materials, they have a wide glass transition temperature range before they reach their melt transition temperature (T_m) and become a liquid. It is within this range that the materials can be formed. By comparison, crystalline and semi-crystalline materials, which have ordered molecular chains, are more difficult to thermoform because they do not have a wide pliable glass transition temperature range and possess only a sharp melt transition temperature value as shown in Figure 4.9.

For example, amorphous materials are similar to a stick of butter. When heated, a solid stick of butter begins to soften. As additional heat is applied, more of the stick softens and some areas begin to melt. Finally, as enough heat is applied the stick melts completely. Crystalline and semi-crystalline materials, on the other hand, are similar to an ice cube. As heat is applied to the ice cube, it begins to melt. The ice cube can only exist as a solid ice cube or

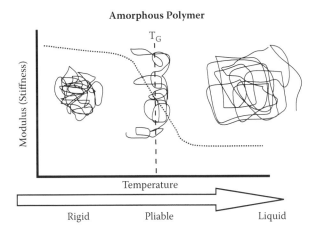

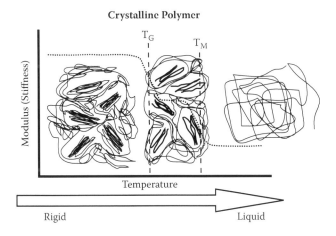

FIGURE 4.9
Amorphous and crystalline material melt temperature transition graphs.

as liquid water. The soft state does not exist. In both cases, the stick of butter and the ice cube started as solids and ended as liquids, but the stick of butter experienced a broad temperature range in which it softened prior to melting completely.

In some applications, materials may require enhanced properties. Common additives include flame retardants, impact modifiers, and UV inhibitors. There are a limited number of additives available for thermoforming materials. The use of additives is not as extensive as in injection molding because a large percentage of the materials formed are clear or semi-opaque since they are used as food packaging containers and closures. Table 4.2 lists some of the more common thermoforming materials along with their properties and some of the main applications.

TABLE 4.2

Common Thermoform Materials

Material	Properties	Sensitivity to Moisture	Applications
ABS	High impact Weather resistant Rigid High strength	Yes	Luggage Enclosures Vehicle parts
HIPS/PS	Low process temperature Fast cycle times Widely used	No	Packaging Toys Low cost applications
PC	Rigid High impact Excellent clarity Heat resistant	Yes	Signs Machine guards Skylights Aircraft parts
PE—HDPE and LDPE	Chemical resistant Translucent Good impact High shrinkage rate	No	Low cost enclosures Vehicle parts
Polyester	FDA approved for food Low process temperature Fast cycle times	No	Food storage Medical device trays Signs
PMMA	High strength Good clarity	Yes	Signs Light covers
PP	Good impact High process temperature Highly chemical resistant	No	Luggage Enclosures Chemical applications Food storage containers Toys
PVC	Good impact Highly chemical resistant Rigid Can be transparent in thinner thicknesses	No	Packaging Vehicle parts

4.6 Thermoforming Part Design Guidelines

Designing parts for the thermoforming process is different from devising other plastic conversion processes for several reasons. As stated previously, the parts are formed from a sheet instead of pellets, powder, or liquid, and they are processed when the material becomes semi-pliable as it is heated. Also unique to this process is the wide variety of materials that can be used to fabricate tooling. Thermoforming can offer fast, inexpensive molds for prototyping. Consult your supplier for specific design requirements.

Wall sections (where t is equal to the nominal wall thickness)

- Material thinning—in male molds, the thinning occurs on the ends and in female molds the thinning occurs toward the middle, as illustrated in Figure 4.10. A good example of thinning is present in thermoformed drinking cups. The base of the cup is thicker than the sides.
- Ribs 3× max high, min 2× material thickness.

Draft

- Male molds—draft is more critical on male molds because as the part cools, it shrinks around the mold and has a more difficult time releasing.
  - Male mold draft of 3°–7°.
- Female molds—draft is not as critical on female molds because as the part cools it tends to shrink away from the mold.
  - Female mold draft of 1°–3°.

Radii

- Minimum radii inside corner of 0.75 t–1 t.
- Avoid sharp edges.

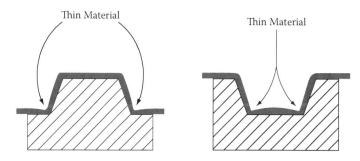

FIGURE 4.10
Design guide illustration: wall sections—areas of material thinning in male and female molds.

Holes

- Holes—drill holes no closer than 0.5″ to 0.625″ (12.7–15.9 mm) from the edge because of notch sensitivity and cracking.

Texture

- Draft texture 1° per 0.001″ (0.025 mm) of texture depth.
- Surface not in contact with the mold typically has a better surface finish.
- Parts can be textured.

Draw ratio

- Ratio of the depth of the part to the width of the part. The ratio depends on the material processed and the process selected. Draw ratios can range from less than 1:1 to 2.5:1 or greater. The part design can also play a roll in maximum ratio.

Part attachment methods

- Friction fit base and cover
- Adhesive
- Foil seals

4.7 How to Identify a Thermoformed Part

In some cases thermoformed parts can be challenging to identify, but for the most part they are thin wall sectioned parts with no gate marks. Below is a list of characteristics that can be used to assist in the identification of a thermoformed part.

- Extremely thin wall sections
- No gate marks
- No ejector pin marks
- Posts and other internal details may be glued in
- No sharp details

4.8 Case Studies

4.8.1 Case Study 1—Fence and Wall Assemblies

The boom of townhouse developments across the United States occurred during the late 1990s and early 2000s. The close proximity of the houses made privacy an important issue for homeowners and the answer was thermoformed privacy fences. The first designs were very simple, one color designs. The sides of the fence were thermoformed and snapped together using formed features. The fence panels were attached at the construction site to extruded corner posts with injection molded decorative covers. Over time, advancements in thermoforming technology permitted textured fences with color to be manufactured. From afar, these fences have the appearance of stone, bricks, etc. U.S. patent 6719277, titled *Thermoformed wall and fence assemblies*, shows examples.

4.8.2 Case Study 2—Polystyrene Clamshell Food Packaging to Thermoformed Packaging

During the 1980s and 1990s a number of U.S. counties and cities banned the use of beaded polystyrene food storage containers.[5] The search for an alternative way to manufacture food containers that were capable of being recycled soon followed. Two-piece thermoformed food containers with a friction fit feature around the perimeter were developed and used in restaurants. Later, one-piece containers with living hinges were developed. Today, multi-colored thermoformed parts with friction fit covers and living hinges are utilized in the food and beverage industry to hold take-out food.

4.8.3 Case Study 3—Form and Fill Food Processing

In the form and fill food processing method, containers are thermoformed using a continuous feed of material. Once the parts have been formed, they are filled and then sealed. Food products such as pudding, yogurt, fruit, and cheese can be processed in this manner. After the container has been sealed it is packaged in boxes and shipped to grocery stores or stored in warehouses.

References

1. R. Stewart, "Thermoforming," *Plastics Engineering*, February 2003.
2. J. Throne, "Understanding How a Sheet Stretches." *Thermoforming Quarterly*, 3rd Quarter, 2005.
3. John Morris, "How It's Made—Thermoforming," Ask an Expert, http://business.articles-and.info (accessed December 11, 2007).
4. Plastic Website, "Thermoforming History," http://www.plasticwebsite.com.au (accessed January 8, 2008).
5. Charles Lake, "Banned," Comfort 'n' Color, http://www.comfortncolor.com/HTML/Ban.html (accessed March 20, 2008).

5

Reaction Injection Molding

Reaction Injection Molding Process Key Characteristics

	Soft Tool	Hard Tool
Volume	Low	High
Material selection	Limited	Limited
Part cost	High	High
Part geometry	Some features	Some features
Part size	Small to very large	Small to very large
Tool cost	Low	High
Cycle time	Minutes	Minutes
Labor	Manual	Automatic

For a list of other conversion process characteristics, see Appendix B.

5.1 Process Overview

Reaction injection molding, or RIM for short, began as a low volume conversion process utilizing thermoset polymers. Recently, it has gained popularity in the automotive markets. Figure 5.1 shows the equipment used in this process involving two reactive chemicals: an isocyanate and a polyol. The two liquids are mixed at pressures between 1500 psi and 3000 psi (10–21 MPa) in the mixing chamber and injected into the mold from the lowest point of the mold in order to minimize trapped air.[1] The part is then allowed to cure before it is removed from the mold. Once the part has been de-molded, it may be necessary to clean the mold before the next cycle. The cured part is then ready for post-molding operations, which include trimming the flash, filling voids, adding inserts, priming, and painting. The filling process of reaction injection molding occurs at very low pressures, around 50–150 psi (0.34–1 MPa), and low flow rates to prevent air entrapment.[2] This process takes place at room temperature and, compared to injection molding, does not require complex tooling to produce parts. Another unique feature of reaction injection molding is that the cycle times are measured in minutes rather than seconds like regular injection molding cycle times.

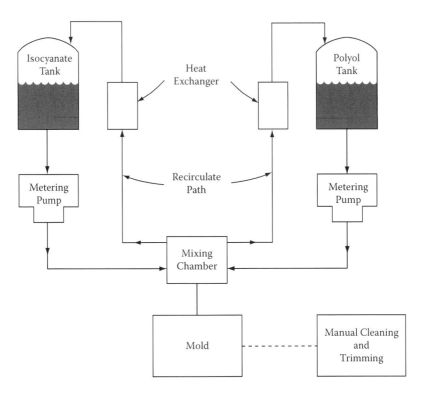

FIGURE 5.1
Diagram of reaction injection molding equipment.

Why would you use this process? It is well suited for very low volume to high volume applications based on tooling, or large parts where metal and other materials are replaced in order to obtain a weight reduction. Although the individual piece part price will be more expensive than a regular injection molded part, in some cases the tooling cost can offset the piece part price in low volumes.

The most familiar applications of this process include

- Automotive (SRIM)
- Agriculture
- Electronics
- Medical equipment housings
- Furniture
- Recreation vehicles
- Sporting goods
- Appliances

5.2 A Brief History of Reaction Injection Molding

Reaction injection molding technology is relatively new compared to other plastic conversion processes. It was developed in Germany in the late 1960s by Bayer AG based on years of polyurethane chemistry experience. Bayer AG first introduced reaction injection molded parts for an experimental "all plastic car" at the plastics fair in Düsseldorf, Germany, in 1967.[3] By the mid-1970s, reaction injection molding had made its way to the United States where it experienced a rapid expansion in use due to the Corporate Average Fuel Economy (CAFE) initiative, which was first enacted by the U.S. Congress in 1975. CAFE was created to reduce energy consumption in cars and light trucks during the energy crisis of the 1970s, and reaction injection molding provided an economical means of producing lighter vehicle fascias and body panels.[4] Later, as automotive manufacturers continued to improve on vehicle weight and fuel efficiency, the process of structural reaction injection molding, or SRIM, was developed. By altering the polyurethane chemistry and/or the preform material, a wide variety of densities and physical properties could be created. In some cases the structural reaction injection molded parts were comparable to their steel counterparts at a fraction of the weight.

Today, RIM and SRIM parts can be seen in optional pickup truck bed liners and agricultural tractor exteriors. Both parts measure about 6′ × 6′ (1.8 × 1.8 m) and weigh close to 100 pounds (45.4 kg) each. Due to the large size and weight of the parts it would be impractical to use conventional injection molding to manufacture these and similar large size parts.

5.3 Equipment

Equipment used in the reaction injection molding process is listed below and is shown in Figure 5.2.

1 – Material feed tanks

2 – Metering pump

3 – Mix head

4 – Heat exchanger

5 – Mold

6 – Manual cleaning tools

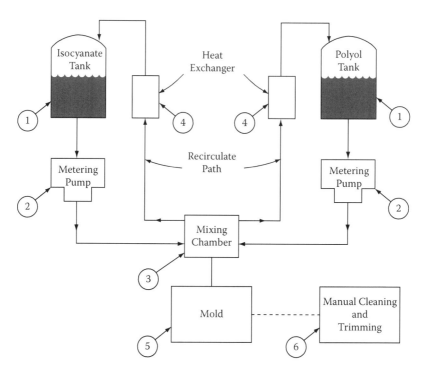

FIGURE 5.2
Reaction injection molding equipment diagram.

5.3.1 Material Feed Tanks

Two tanks feed the primary materials, polyol and isocyanate, to the mixing head where the materials react to form a thermoset polyurethane plastic. Tank pressures are typically 40 psi (0.28 MPa) or less.

5.3.2 Metering Pump

Metering pumps are used for precise delivery of the correct ratio of polyol and isocyanate to the mix head. Material properties such as density, flexural modulus, and impact strength can be altered by changing the mix ratio of the materials.

5.3.3 Heat Exchanger

Heat exchangers are used to maintain the temperature of the polyol and the isocyanate. This is necessary to control viscosity and facilitate homogeneous mixing.

5.3.4 Mix Head

The function of the mix head is to homogeneously combine the polyol and isocyanate prior to injecting the material into the mold. This is done by impinging or directing the materials against each other at high speeds, which results in a turbulent flow. Between cycles, material that has not been mixed is recirculated through the system.

5.3.5 Mold

The mold is shaped to match the contours of the part that is to be molded. It is similar to a conventional injection mold except it lacks a cooling system and an ejection system. Mixed thermoset materials are injected into the mold at low pressures. This is done to maintain laminar flow of the material, which helps to minimize air bubbles. The part is allowed to cool and then it is removed from the mold. The silicone mold shown in Figure 5.3 has a bottom half and a top half.

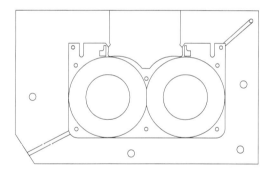

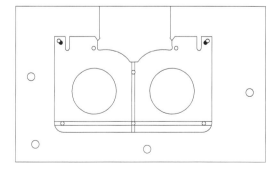

FIGURE 5.3
Two-piece reaction injection mold.

5.3.6 Manual Cleaning Tools

Tools used to clean the mold and parts include knives, scrapers, files, and abrasive media. The cleaning tools are listed to emphasize that the reaction injection molding process needs manual cleaning. This is not a "shoot and ship" process.

5.3.7 Post Molding

Once a part has been molded it will require some type of secondary operations or finishing work to complete the part. Molds are designed to allow overflow or flash to ensure part fill or completeness, and this will have to be removed from each part after it has cured. Also inherent to this process are small bubbles or voids. On cosmetic surfaces, bubbles or voids will need to be filled with a two-part plastic body filler, allowed to cure, and the excess material removed using an abrasive media. Finally, primer and paint are applied where required. Other finishing procedures may include removing the material between the slots of a part designed to have a venting area, for example, a computer monitor housing. Again, it is important to note that all reaction injection molded parts require some sort of finishing work before they are complete. This can be a labor intensive process and is one reason why the piece part prices are considerably higher when using this process.

5.4 Tooling

5.4.1 Mold Materials

For larger volume reaction injection molded parts, molds are typically fabricated from aluminum or steel similar to other molding processes, but for smaller quantity runs (under 50 pieces) silicone can be used to create the molds.

5.4.2 Mold Fabrication

Molds used in the reaction injection molding process have five rather distinct characteristics: (1) the tooling is designed such that the cosmetic surface of the part is molded upside down so any trapped air bubbles form on a non-cosmetic surface; (2) the mold is mounted in the press at an angle to minimize the possibility of air being trapped inside the mold; (3) flash is designed into each mold to confirm the mold has been filled completely; (4) the gate is located on the lowest end of the mold; and (5) the vents are on the upper end to allow air to evacuate the mold cavity. Figure 5.4 shows a cross-sectional view of a mold.

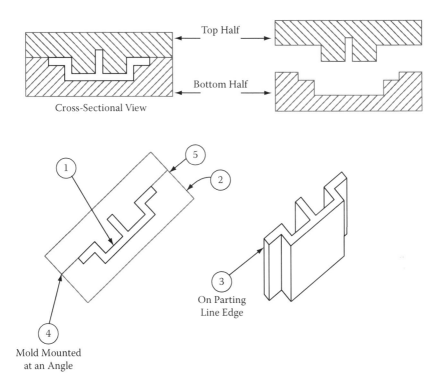

FIGURE 5.4
Cross-sectional view and a tilted view of a reaction injection mold.

As described previously, silicone molds, which can be used for low volume, prototype, or quick turnaround tooling, are usually created using rapid prototyping technology. A stereolithography (SLA) model is produced to act as the pattern. The next step involves the creation of a silicone mold using a standard process within the industry. The silicone is poured into a rectangular base and the pattern is placed on top of the silicone and pressed into place to create a logical parting line. After the silicone has hardened, a release agent is applied to the exposed surface, which will later become the parting line of the mold. The other half is created by pouring additional silicone over the pattern and allowing it to cure. The mold halves are separated and the pattern is removed leaving a void, which is later filled during the reaction injection molding process. The pattern is saved so it can be used to create additional molds if necessary.

For higher volumes, molds are constructed from aluminum with cartridge heaters or heating lines. Aluminum or steel tooling is machined to match the contours of the molded parts. The gate and vents are added as required. In most cases, an edge gate is used to deliver the resin mixture to the mold cavity. Reaction injection molding presses have the ability to tilt the molds so the material can enter the mold cavity in a laminar flow using low pressure and low temperature. As previously stated, laminar flow is important because it minimizes air bubbles that can become trapped in the mold, which create voids after the part has cured. Since this process uses heat to cure the materials, cooling is not required. This helps to reduce the cost of reaction injection molding tools and processing compared to other plastic conversion processes where cooling is required.

5.5 Materials

Materials in this process are delivered to the molder as one of two components, either an isocyanate or polyol, both of which are liquids. When these two liquids mix a reaction between them occurs to form a polyurethane thermoset. Once the mixture is allowed to cure, the resulting material properties are most similar to the thermoplastic ABS (acrylonitrile butadiene styrene). Like conventional injection molding materials, reaction injection molding materials can include additives and/or fillers. Although the number of reaction injection molding additives and fillers is only a fraction of the number available to conventional injection molding, there are common property enhancing additives and fillers that change flexural modulus, density, or impact strength.

During the molding stage of the process the reaction rate of the isocyanate and polyol must be carefully controlled; if the reaction rate is too fast, the material begins to solidify and the mold will not fill completely.[5] Also if a part is de-molded too early during the curing stage, the overall dimensions can be affected and the part can be easily damaged.

5.6 Reaction Injection Molding Design Guidelines

Parts designed to take advantage of the RIM process should follow these basic guidelines. You should always check with your molder for exact requirements and capabilities. Part flash or overflow is required to confirm that the molded part has completely filled.

Wall sections (where t is equal to the nominal wall thickness)
- Maintain an optimum uniform wall thickness ranging from 0.06″–0.30″ (1.5–7.6 mm) as illustrated in Figure 5.5.
- Maximum wall thickness of 1.25″ (31.75 mm).

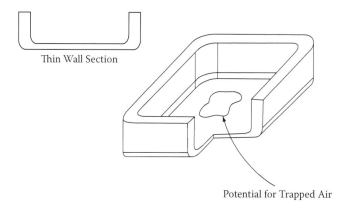

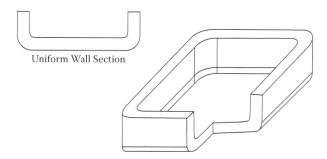

FIGURE 5.5
Design guide illustration: wall sections—thin wall sections compared to uniform wall sections.

- Minimum of a 1° draft angle; 2° is preferred to assist in the demolding process and this also helps to minimize surface damage to the part as shown in Figure 5.6.
- Extreme variations in wall sections without a transition may result in difficulty filling the part and trapped air or voids may be present in the thin wall sections.
- Maintain a constant wall section on corner bosses as shown in Figure 5.7.

Radii

- Minimum radius of 0.06″ (1.5 mm).
- No sharp edges.

Ribs

- Should be in the direction of material flow. Since this is a low pressure process parts are not "packed" out; as they shrink

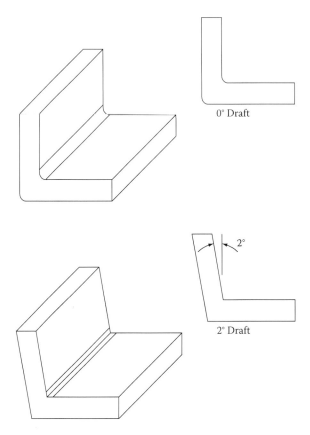

0° Draft

2°

2° Draft

FIGURE 5.6
Design guide illustration: wall sections—0° draft angle compared to a 2° draft angle.

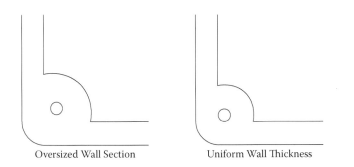

Oversized Wall Section　　　　Uniform Wall Thickness

FIGURE 5.7
Design guide illustration: wall sections—corner boss.

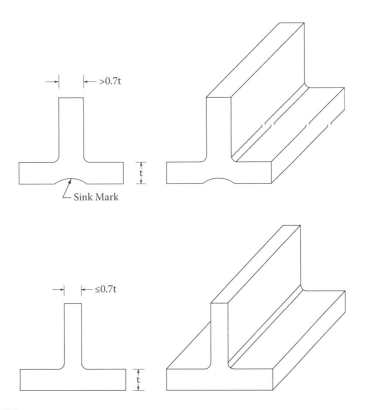

FIGURE 5.8
Design guide illustration: ribs—ribs greater than 70% of the nominal wall thickness compared to ribs less than 70% of the nominal wall thickness.

during the cycle there is a risk that air will be trapped when the two flow fronts meet inside the mold.

- Single rib designs, illustrated in Figure 5.8, should be a maximum thickness of 0.7 t of the mating wall section or sink may appear on the part.
- Ribs exceeding 0.7 t should be converted to multiple rib designs as shown in Figure 5.9. A minimum spacing of 1 t is recommended between ribs.
- Multi-directional rib designs as shown in Figure 5.10 tend to result in trapped pockets of air, which result in voids that may require additional secondary operations to complete the part after it is de-molded. The location of the air pockets can vary, depending on the gate location.

Gussets

- Maximum wall thickness up to 0.7 t of the thickness of the adjoining wall.

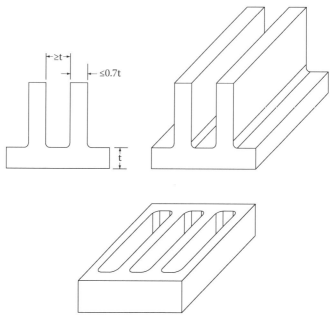

FIGURE 5.9
Design guide illustration: ribs—thick ribs can be converted to multiple ribs.

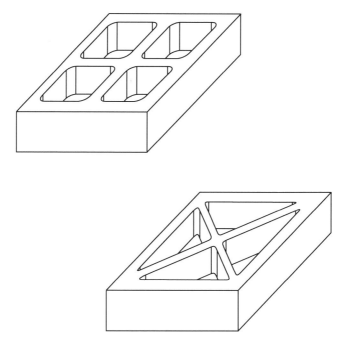

FIGURE 5.10
Design guide illustration: ribs—multidirectional ribs.

- To ensure that the feature is properly filled and vented, the gussets should be in the direction of material flow since parts in this process are not "packed" out during the cycle. An example is shown in Figure 5.11. Note that since this is a low pressure process, these gussets differ from injection molded gussets. They have a longer length in contact with the nominal wall section. There is a risk that air will be trapped when the two flow fronts meet inside the mold.

Boss

- Maximum wall thickness of 0.75 t of the nominal wall thickness.
- Add gussets in the direction of material flow to assist in filling out the part detail. This helps to minimize trapped air when the two flow fronts meet inside the mold.

Holes

- Minimum of a 1° draft angle; 2° is preferred to assist in the demolding process.
- Holes not in the direction of pull can be created with pick-out cores.
- Holes with no draft can be drilled as a secondary operation.

Part attachment methods

- Overlapping step joints.
- Snap fits—can be designed similar to injection molding snap fits.
- Adhesive.
- Fasteners and inserts can be installed using ultrasonic welding.

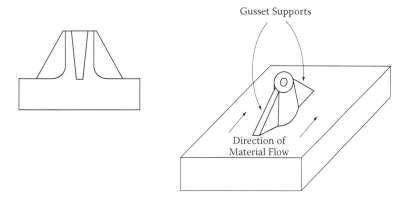

FIGURE 5.11
Design guide illustration: gussets—boss with gussets.

5.7 How to Identify a Reaction Injection Molded Part

Below is a list of characteristics that can be used to assist in the identification of a reaction injection molded part.

- Walls sections up to 1.25″ (31.75 mm)
- Thick wall sections
- Flash designed into the mold to confirm part fill
- Part needs to be painted
- Polyurethane thermoset
- Bubbles and voids
- No injector pin marks on B-side
- Limited internal detail
- Lack of sink marks on adjoining wall sections

5.8 Case Study

5.8.1 Case Study 1—Fast Prototypes That Function

Although reaction injection molding can be used to produce low volume parts it can also be used to make prototype parts. In this particular case, it was chosen because multiple parts were needed and stereolithography technology was not developed at the time.

Due to a tight timeline, reaction injection molding was selected because multiple molded parts could be delivered within a week. Had injection molding been selected for prototype samples, the lead times would have been over 12 weeks at a much higher cost. Ten sample assemblies were produced and distributed to the project team and the customer. This provided an excellent opportunity to review and evaluate the design because the parts were dimensionally accurate and the critical product features were present. These assemblies signaled to the customer that the project team had completed the conceptual design stage and was ready to proceed with long run injection molds. Reaction injection molded parts also provided information on fit and function without expensive and time-consuming tooling modifications compared to an aluminum or steel mold. The time it took to go from concept to launch was 8 months as opposed to 12–16 months due to machining times.

References

1. Harry George, "RIM Goes Bumper to Bumper," Machine Design, http://machinedesign.com (accessed November 18, 2007).
2. Bay Systems, "Advantages of RIM," http://www.bayer-baysystems.com (accessed February 4, 2008).
3. Bayer Group, "Auto Creative—Innovation News," http://www.bayer-materialscience.com (accessed December 7, 2007).
4. United States Department of Transportation, "Laws/Regulations/Guidance," National Highway Traffic Safety Administration, http://www.nhtsa.dot.gov (accessed August 8, 2008).
5. SRI International, "The Role of NSF's Support of Engineering Technological Innovation," http://www.sri.com (accessed February 4, 2008).

6

Rotational Molding

Rotational Molding Process Key Characteristics

Volume	Medium
Material selection	Limited
Part cost	Medium to high
Part geometry	Simple
Part size	Small to very large
Tool cost	Low to medium
Cycle time	Minutes
Labor	Manual

For a list of other conversion process characteristics, see Appendix B.

6.1 Process Overview

Rotational molding, which is commonly referred to as rotomolding, is a plastic conversion process for molding large, hollow stress-free parts. This process, unlike other plastic conversion processes, does not use external pressure to mold parts, only heat. Resin, in the form of a fine powder or a liquid, is placed inside a mold base. The cover is then placed over the base and secured into position. The mold is transferred to the oven where it is rotated about two perpendicular axes at independent speeds of less than 20 RPM.[1] As heat is applied, the material becomes tacky and joins together in a process called fusion. The rotation of the mold allows the sides to be evenly coated as the material is heated until fusion occurs.

Once the materials have completely melted, the mold is transferred to the cooling stage where air or water is used to remove the heat from the mold. In the last step, the mold is opened and the part is removed. Cycle times for this process are measured in minutes rather than seconds. Some cycle times can be 30 minutes or more. The rotational molding process is shown in Figure 6.1.

Why would you use this process? It is well suited for very large, hollow parts that would be impractical to mold using injection molding or blow molding. Rotational molded parts can be seen in agricultural liquid storage tanks, automobile side panels, boat hulls, playground equipment, and trash cans.

The most familiar applications of this process include

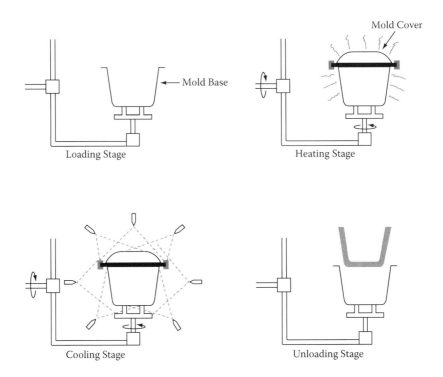

FIGURE 6.1
Diagram of the rotational molding process.

- Agriculture
- Automotive
- Housewares
- Industrial (storage drums)
- Sports and leisure
- Transportation and traffic safety
- Toys

6.1.1 Variations of the Rotational Molding Process

There are four variations of equipment that are used in this process, but they ultimately function the same way. The first is the "carousel configuration," which is illustrated in Figure 6.2. Multiple molds are mounted to rotating arms and transferred among three positions: the load and unload position, the heating position, and the cooling position. Molds are loaded with materials and heated in an oven. Once fusion has occurred, the mold is removed from the oven and allowed to cool. The part is removed and the mold is prepared for the

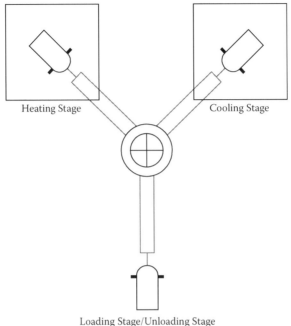

Loading Stage/Unloading Stage

Carousel Configuration

FIGURE 6.2
Variations of the process—carousel configuration.

next cycle. This configuration yields the most parts per hour and is very energy efficient, but the major drawback is that it has a large footprint.

The "shuttle configuration" is shown in Figure 6.3, and is the second variation of the rotational molding process. The mold is transferred to and from the oven via a cart. Multiple carts can be used in this process. Molds are loaded and unloaded at one station, similar to the carousel configuration, and the cart is moved into the oven where it is heated. The cart is removed from the oven once fusion has occurred. The next cart is shuttled into the oven and the previous cart is allowed to cool. Once cooled, the part is removed and the mold is prepared for the next cycle. This variation can consume a large amount of floor space, especially since the shuttle system is located on two sides of the oven.

The third variation is a "clamshell configuration" and features a mold with a hinged cover mounted to an arm. Figure 6.4 shows this variation. The loading, heating, cooling, and unloading all occur at the same station. After the mold is loaded with material, the mold cover closes over the base and is secured in place. As with the previous two variations, the next step is to heat the mold, causing fusion, and then allowing it to cool. The part is

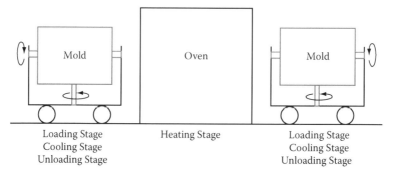

Shuttle Configuration

FIGURE 6.3
Variations of the process—shuttle configuration.

removed and the mold is set up for the next cycle. Although the clamshell style equipment is a low capital investment and takes up a smaller footprint, the throughput is limited because the process occurs in series with a single mold station versus multiple molds.

The fourth configuration, illustrated in Figure 6.5, is the "rock and roll configuration" because the mold is not rotated about two axes as with other variations. Instead, the mold is "rocked" back and forth in one direction, while it "rolls" or rotates in a perpendicular direction. This configuration is used for long, narrow parts such as kayaks and playground equipment.

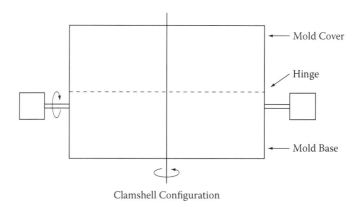

Clamshell Configuration

FIGURE 6.4
Variations of the process—clamshell configuration.

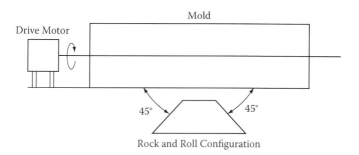

Rock and Roll Configuration

FIGURE 6.5
Variations of the process—rock and roll configuration.

6.2 A Brief History of Rotational Molding

Rotational molding is a process that most closely resembles the casting process used to make porcelain and ceramic articles in the 17th century. Both methods rely upon a buildup of material along the walls of a mold. After a period of time, the mold halves are separated and the part is removed. While there is no evidence of a direct connection, it seems logical that the idea of casting a plastic part could have its origins long ago.

The first use of rotational molding with plastics occurred in the United States in the late 1940s. The toy industry made use of this process to mold doll heads using PVC. This process was then introduced in Europe in 1953 where it was initially used in the toy industry as well.[2] Uses and applications for rotational molded parts began to grow slowly with the introduction of traffic safety cones, but the limiting factor in this process was based on the materials available for processing.

Polyethylene was introduced in rotational molded parts in the 1960s. For the first time, large liquid storage containers could be molded economically using this method. Almost 50 years later, this process is still used for the production of large storage drums and containers. Later, additional materials became available, such as linear low density polyethylene, polypropylene, and nylon, which fueled further growth in this field.

Over the past few decades, the focus of this process has been on improving the cycle times and piece part quality. Advances in equipment and processing technology now allow the operators to better control the temperature within the oven so the part is heated to the correct processing temperature, as well as when the part has cooled significantly to be removed from the mold. As the cycle times decrease, so does the effective piece part price, which helps this process to better compete with other conversion processes. This has also helped to increase the range of rotationally molded part applications.

Today, polyethylene is the primary material used, but a number of other materials are available. The uses of this process continue to grow.

6.3 Equipment

Equipment used in the rotational molding process is listed below and a sketch was shown previously in Figure 6.1. The equipment has been grouped into stages of operation to simplify the explanation.

1 – Loading
2 – Heating
3 – Cooling
4 – Unloading

6.3.1 Loading Stage

The mold is mounted to an arm, which is connected to a motor. This enables the mold to be rotated. This assembly is either attached to the main piece of equipment or to a movable cart. When in this position, a precise amount of powder or liquid material is weighed and placed into the mold. The mold is closed and the cover is secured to the base by means of a clamping system. From here, the mold enters the heating stage. The thickness of the part can be changed, depending on the amount of material placed in the mold.

6.3.2 Heating Stage

Once the mold has entered the oven it is heated as it rotates. Blowers or fans are used to evenly distribute the heat inside the oven. Depending on the materials processed, the temperatures range from 500°F–700°F (260°C–371°C).[3] The mold rotates as the material melts and coats the sides of the mold evenly. Proper heating of the mold is important as it can affect the properties of the final product. Figure 6.6 illustrates various stages of coating the mold as it is heated. After fusion has occurred, the mold is ready to be cooled. The heating stage is the longest step in the rotational molding process, but this gives the operator enough time to prepare a second mold for the next cycle.

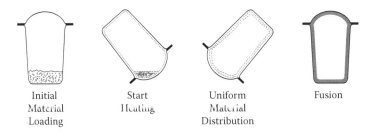

Initial Material Loading | Start Heating | Uniform Material Distribution | Fusion

FIGURE 6.6
Various stages of how the material coats the mold as it is heated.

6.3.3 Cooling Stage

Since rotational molds do not have integrated cooling lines, like injection molds, forced air or water jets are sprayed on the outside surface of the mold in order to cool it. This is done to help maintain a constant rate of cooling throughout the entire part. If the part is cooled too quickly, the part may shrink rapidly, which could cause it to warp.

6.3.4 Unloading Stage

After the mold has cooled, the cover is unclamped and removed. Depending on the size of the part, it is removed either manually or by a machine. Since the heating stage is the longest part of the cycle, there is enough time to prepare a second mold for the next cycle. This includes cleaning the mold, measuring out materials, loading the materials into the mold, and clamping the mold closed. The operator also has a sufficient amount of time to do any finishing work to the newly molded part if required.

6.4 Tooling

6.4.1 Mold Materials

Molds can be fabricated from sheet metal or solid plates. In most cases, sheets of aluminum or steel are used to fabricate rotational molds. There are two main reasons molds can be made from sheet metal versus solid plates. The first is that this is a zero pressure conversion process, so the mold only needs to support its own weight and the weight of the resin. The second reason is that thinner mold walls allow the heat to transfer faster through the mold, into the material. It also allows for consistent cooling of the mold. Solid aluminum plates can also be used to fabricate molds, but they are used for higher volume tools that feature complex designs or very intricate detail.

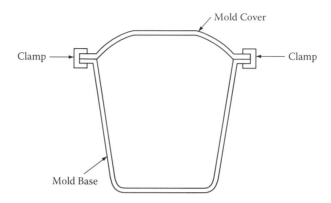

FIGURE 6.7
Example of a rotational mold.

6.4.2 Mold Fabrication

Molds used in this process are very thin compared to molds used in other processes. The majority of molds are constructed in two pieces, featuring a base and a cover. The aluminum or steel sheets are bent and formed to the desired shape. Figure 6.7 shows a sheet metal mold with clamps to secure the cover to the base. These molds are simple in design and have an average surface finish. If a more detailed or a higher quality surface finish is required, machined or cast aluminum molds can be used. Cast aluminum molds require a pattern part to create the mold and can make this type of mold the most expensive of the three fabrication methods.

Some part designs may require undercuts. They are difficult but not impossible to mold using this process. Multiple-piece molds can be used to accomplish this, but it adds to the tooling costs and associated labor content. Complex part designs may also take advantage of the multiple-piece construction to create difficult to form features.

Vents are required in the mold to prevent the buildup of pressure on the inside of the mold as the material is heated. Although no pressure is used to form the part, pressure can develop as a result of the heating process. Excess pressure can deform the mold and cause the finished part to warp when it has cooled and been removed from the mold. The diameter of the vent is directly related to the volume of the part. Multiple vents can be used for very large or complex parts.

Some parts require a large opening on one side. This can be accomplished by creating insulated areas within the mold, which do not allow the transfer of heat to the resin. A container with no cover is illustrated in Figure 6.8. If parts require holes, in some cases they can be created using a button design, which is further explained in the design guide of this chapter.

Finally, rotational molds do not need to be modified in order to change the wall thickness of the part. This is similar to thermoforming where the part thickness can be modified by altering the thickness of the material. Since

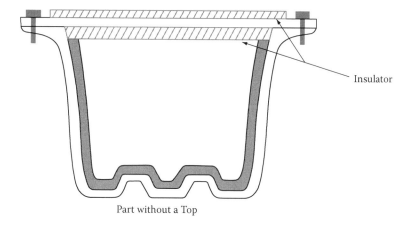

Part without a Top

FIGURE 6.8
Mold designed to create a container with no cover.

tooling is relatively inexpensive, this process can also be used to prototype parts that will ultimately use other molding processes. For example, rotational molding can be used to prototype blow molded parts.

6.5 Materials

Although some liquid formulations exist, the materials used in this process are typically supplied in the form of a fine powder, with particle sizes varying from around 0.0059″ 0.0197″ (150–500 microns).[4] Rotational molding has a very limited number of materials available for processing. Currently, about 85% of all parts use polyethylene, which is an inexpensive commodity-grade resin with good impact qualities. Polyvinyl chloride is the second most used material in this process at about 10%. Acrylonitrile butadiene styrene, nylon, polypropylene, and a few others make up the remaining materials. Table 6.1 lists some of the more common rotational molding materials along with their properties and some of the main applications.

The number of materials available for processing is limited for several reasons. First, the materials must resist degradation due to long heat cycles; second, the supply of high temperature ovens that are required for proper fusion of the material is low; finally, the range of product applications is small.

Antioxidants are also added to rotational molding materials to help minimize material degradation during the long heating cycle. Depending upon the material and concentration, the addition of fillers can inhibit fusion of resin molecules.

TABLE 6.1

Common Rotational Molding Materials

Material	Properties	Sensitivity to Moisture	Applications
PE—HDPE and LDPE	Chemical resistant Translucent Good impact High shrinkage rate	No	Chemical storage tanks Fuel tanks Vehicle parts Trash containers Playground equipment Pallets
PVC	Good impact Highly chemical resistant Rigid	No	Sealed bladders Inflatable components
ABS	High impact Weather resistant Rigid	Yes	Storage containers
Polyamide	Chemical resistant Translucent Good impact High temperature	No	Tanks Shrouds
PP	Good impact Highly chemical resistant	No	Storage tanks

6.6 Rotational Molding Part Design Guidelines

Design guidelines for rotational molded parts are not as extensive as those for other processes. The parts tend to be simple shapes and larger than average molded parts. Undercuts can be molded, but should be avoided where possible. Working with your molder will help to ensure that the part can be molded to meet your specifications.

Wall sections (where t is equal to the nominal wall thickness)

- Uniform wall thickness due to the coating process, but outside corners are usually thicker than the other wall sections of the part.
- Maintain an optimum uniform wall thickness ranging from 0.06″–0.50″ (1.52–12.70 mm).
- Maximum wall thickness of 1.2″ (30.5 mm).
- Minimum of a 2° draft angle.
- Double wall parts should have a minimum of five times the nominal wall thickness between walls, as shown in Figure 6.9.
- Avoid large flat areas by adding ribs.

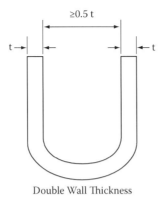

Double Wall Thickness

FIGURE 6.9
Design guide illustration: wall sections—double wall.

Radii

- Minimum radius of five times the nominal wall thickness, as shown in Figure 6.10.
- No sharp edges.

Ribs

- Single wall ribs are ideal for injection molding and compression molding, but are very difficult to mold using rotational molding. Expanding the rib allows it to be rotationally molded. Figure 6.11 shows the difference between a single wall rib and an expanded wall rib. Also shown is the multiple rib design.

Holes

- Holes are typically created by forming a button on the part and cutting it off in a secondary operation. Figure 6.12 shows an example of a button.

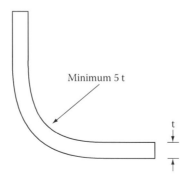

FIGURE 6.10
Design guide illustration: radii—minimum preferred radius.

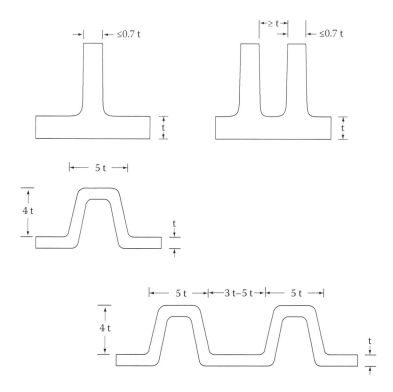

FIGURE 6.11
Design guide illustration: ribs—comparison of single wall injection molded ribs versus expanded wall rotational molded ribs.

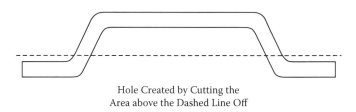

Hole Created by Cutting the
Area above the Dashed Line Off

FIGURE 6.12
Design guide illustration: holes—secondary operation can be utilized to create a hole by removing the button.

6.7 How to Identify a Rotational Molded Part

Rotational molded parts can be identified by some of the characteristics listed below. These parts come in a wide range of sizes and in some cases they can be very large.

- Uniform wall thickness
- Hollow
- Simple geometry
- Large size
- Non- glossy surface finish

6.8 Case Studies

6.8.1 Case Study 1—Traffic Construction Cones

Construction zones are required to be clearly marked to protect the job site workers and motorists. To satisfy this requirement, large safety cones were designed, making them some of the first rotational molded parts to appear in large volumes. To eliminate the need to add additional weight, the cones are molded with thick wall sections. This provides stability and ensures that the cones will withstand high winds, environmental climate changes, and passing vehicles. To further increase the visibility of the cones, reflective markings were added around the circumference of the cone. Today millions of these cones are in use on roadways all over the world.

6.8.2 Case Study 2—Water Tank (Base and Cover)

Large liquid storage containers can be economically produced using the rotational molding process. These tanks are typically used to hold water and can be seen in the agricultural industry. Previously, agricultural water needs were met with troughs or small barrels. This often involved a lot of time checking containers for water levels and handling the equipment required to fill the containers. The plastic rotational molded tanks are translucent and provide the ability to view the current water levels even at a distance.

Some small towns across the United States use large liquid storage containers to water trees and flowers. The large water tanks are placed in a pickup truck bed and filled with water, eliminating the use of a special tanker truck to water the areas.

References

1. D&M Plastics, Inc., "The Rotational Moulding Tehnique," Plastic Moulding, http://www.plasticmoulding.ca (accessed June 6, 2008).
2. N. Ward, "A History of Rotational Moulding," Plastiquarian, http://www.plastiquarian.com (accessed October 8, 2007).
3. D&M Plastics, Inc., "The Rotational Moulding Tehnique," Plastic Moulding, http://www.plasticmoulding.ca (accessed June 6, 2008).
4. Association of Rotational Moulders Australia, "Materials for Rotational Moulding," Rotational Moulding, http://www.rotationalmoulding.com (accessed April 20, 2008).

7

Compression Molding

Compression Molding Process Key Characteristics

	Vertical	Transfer
Volume	Medium	High
Material selection	Limited	Limited
Part cost	Medium	Low
Part geometry	Simple	Some features
Part size	Small to large	Small to large
Tool cost	Medium	High
Cycle time	Minutes	Minutes
Labor	Manual	Manual

For a list of other conversion process characteristics, see Appendix B.

7.1 Process Overview

Compression molding is a moderate to high volume, high pressure thermoset conversion process capable of molding a range of simple parts with superior strength. This part of the process is most similar to plug assist thermoforming with the exception that the forming process uses a hydraulic ram rather than vacuum. Other aspects of compression molding resemble injection molding. In some cases, thermoplastic material can be molded, but the predominant materials in this process are thermosets. Material, referred to as a charge, is placed between the mold halves. The mold closes and pressure is applied, filling the cavity with material. As the pressure is applied, the material is heated, which causes it to cure. Once the part has cured, the mold opens and the part is removed. A compression molding diagram is shown in Figure 7.1. This process is the oldest and most common method of forming thermoset materials.[1]

Why would you use this process? This process is well suited for processing thermoset materials. It is used for products requiring either high strength and/or high temperatures.

The most familiar applications of this process include

- Electrical enclosures
- Automotive parts

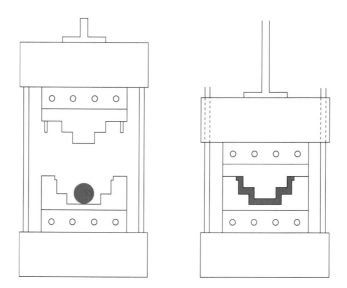

FIGURE 7.1
Diagram of compression molding equipment.

7.1.1 Variations of the Compression Molding Process

Transfer molding, illustrated in Figure 7.2, is a variation of compression molding that resembles injection molding in some aspects. Thermoset material is placed into a transfer chamber where it is heated and a plunger forces the material through a runner and into the mold. After the part has cured, the mold is opened and the part is ejected. One drawback to this process is that the runner is still attached to the part. This will have to be removed and since it is a thermoset material, it cannot be reused in this process.

7.2 A Brief History of Compression Molding

Compression molding was first accomplished in 1907 by Leo Baekeland using phenol-formaldehyde resin. This equipment was primitive and remained so for another 20 years. Eventually, a crude automatic compression machine was developed and a new process was born. In honor of the creator, the phenol-formaldehyde resin was called Bakelite. It produced very hard but brittle products.

The invention of the automobile and its need for tires spurred companies like Goodyear and Goodrich to invest heavily in compression molding processes.[2] Although no longer in the tire business, in 1946 Goodrich invented the inflatable tire design we enjoy today.[3] Gradually, more materials were

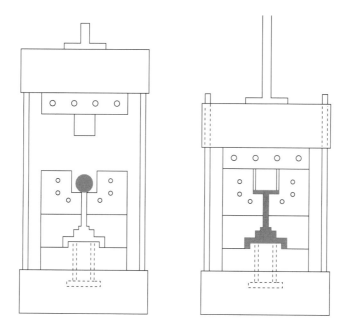

FIGURE 7.2
Variations of the process—transfer molding.

developed and the equipment has been improved. The primary advantage of this process is that the materials can be heavily loaded with reinforcements, such as long glass fibers. Material choices include phenolics, urea-formaldehyde (melamine) mixtures, silicones, epoxies, and polyester, among others.[4]

7.3 Equipment

Equipment used in the compression molding process is listed below and shown in Figure 7.3.

1 – Hydraulic ram
2 – Heated platens
3 – Plunger (transfer molding)
4 – Mold
5 – Ejector assembly
6 – Base

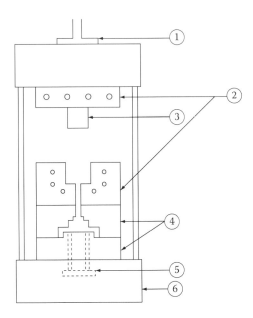

FIGURE 7.3
Transfer molding equipment diagram (with arrow callouts for each stage).

7.3.1 Hydraulic Ram

Pressure is applied to the mold via the hydraulic ram. As the mold halves are closed, the material is forced to fill the cavity. Pressure is maintained as heat is applied to the mold.

7.3.2 Heated Platens

Heated platens are attached to both sides of the press and the molds are bolted to the platens. The platens deliver heat to the mold through a series of heater cartridges. Similar to an injection molding press, a series of slide rods allow one half of the mold assembly to move up and down. The other half is stationary.

7.3.3 Plunger

The plunger is attached to the upper platen of the press and moves downward, forcing the material charge to fill the sprue and mold cavity.

7.3.4 Mold

Compression molds are most similar to injection molds in that they feature two mold halves with a logical parting line. Material is placed between the mold halves and as the compression side of the mold closes, it comes in contact with the molding compound. The material is forced into the voids of the cavity, under high pressure, and as the heat is applied, the material cures.

7.3.5 Ejector Assembly

The ejector assemblies of compression mold are comparable to injection mold ejector assemblies. They serve to assist in the ejection of the part from the mold once it has cured.

7.3.6 Base

The base or stationary side of the compression molding press contains half of the mold and the ejector assembly. Material is placed in the mold on this half of the press prior to the mold closing.

7.4 Tooling

7.4.1 Mold Materials

Similar to other conversion processes, compression molds are fabricated using two or more plates of steel or stainless steel. These materials are used for their ability to transfer heat and withstand the high pressure exerted during the cycle.

7.4.2 Mold Fabrication

Compression molds are fabricated using the same methods as the other conversion processes, namely, machining or the EDM process. These molds do not feature complex geometries or fine detail so they can be designed and built in a relatively short period of time. They are similar to two-plate injection molds in that a portion of the part detail is contained on each side of the mold. Transfer molds are three-piece assemblies: the two mold halves and the third piece featuring a material charge chamber and a runner system. The lower half of the compression mold is typically the female half and the material charge is placed in the cavity of the mold prior to the mold closing. In a transfer mold, the lower half of the mold can be either a male or female mold since the material charge is placed in the chamber above. Molds have a flash trap or overflow feature, which is used to ensure that the part is filled completely.

7.5 Materials

Thermoset materials are used to mold the parts in the compression molding process. Heat is used to cure the material where cooling is utilized to solidify the parts in other processes such as injection molding, extrusion, blow molding, and thermoforming. Material is processed in one of two forms, either as a bulk molding compound (BMC) or sheet molding compound (SMC). Bulk molding compounds are a mixture of a base resin, additives, and fillers that when combined form a pliable substance that resembles putty. Sheet molding compounds, on the other hand, come in rolls and are cut to the desired size based on the size of the part being molded. One advantage of this process is that the charge is based on weight, which minimizes the amount of waste generated during each cycle.[5] Since this material is a thermoset, reject parts cannot be ground up and reprocessed, so obtaining the optimal weight is essential to molding parts cost effectively. It is also important to note that, compared to thermoplastic materials, some thermosets have a limited shelf life before processing.

7.6 Compression Molding Part Design Guidelines

Parts designed for this process should follow these basic guidelines. Compression molded parts tend to have less rigorous guidelines than injection molded parts. The molder can assist with design specifications.

Wall sections (where t is equal to the nominal wall thickness)
- Maintain an optimum uniform wall thickness ranging from 0.04″–0.30″ (1.02–7.62 mm).
- Extreme wall thickness variations possible. Since this is a manual loading process, additional materials can be placed where needed.
- Minimum of a 1° draft angle [1° = 0.017″ per inch (1° = 0.43 mm per mm)] required to eject the part.
- Curing time is directly related to wall thickness.

Texture
- Minimum draft texture 1° per 0.001″ (0.025 mm) of texture depth.

Radii
- Minimum radius of 0.3 t of the nominal wall thickness.
- Minimum radius of 0.03″ (0.76 mm) with 0.06″ (1.5 mm) preferred.
- No sharp edges.

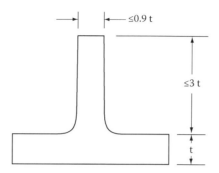

FIGURE 7.4
Design guide—single rib design.

Ribs

- Single rib designs, illustrated in Figure 7.4, should be a maximum thickness of 0.9 t of the mating wall section and have a maximum height of 3 t.
- Ribs exceeding 0.9 t wide should be converted to multiple rib designs, as shown in Figure 7.5. A minimum spacing of 2 t is recommended between ribs and a maximum height of 3 t.
- The ribs can vary in height and in width as shown in Figure 7.6, but should follow the guidelines above. Additional material can be placed at the specific location to mold the ribs.

Boss

- Height to be maximum of 2 d compared to the diameter, as shown in Figure 7.7.
- Minimum of a 4° draft angle [1° = 0.017″ per inch (1° = 0.43 mm per mm)] required to eject the part; 5° draft angle preferred.

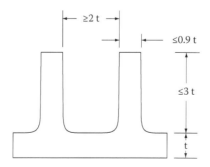

FIGURE 7.5
Design guide—multiple rib design.

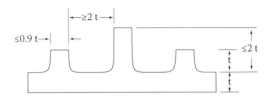

FIGURE 7.6
Design guide—variable rib design.

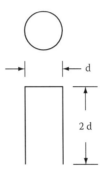

FIGURE 7.7
Design guide—boss height.

Holes

- Minimum molded hole size is 0.06″ (1.5 mm); smaller holes need to be drilled as a secondary operation.
- Minimum of a 1° draft angle; 2° preferred to assist in part ejection.
- No closer than 2 t from edges.
- Maintain a minimum spacing of 2 t between holes.

Flash

- Designed into parts to confirm part fill.
- Compression molding typically less than 0.004″ (0.102 mm).
- Transfer molding typically less than 0.005″ (0.127 mm).

Part attachment methods

- Fasteners and inserts can be installed using ultrasonic welding.

7.7 How to Identify a Compression Molded Part

Most simple parts that are compression molded are made from thermoset materials. Some parts may have signs of a gate, depending on the process.

The parts can range in size from small to large, and since flash is inherent to the process, traces of where the flash was trimmed may be visible. Additional characteristics follow:

- Simple geometry
- Rigid parts
- Heavy parts

7.8 Case Study

7.8.1 Case Study 1—Electrical Service Box

A traffic monitoring system originally used a powder-coated metal box to house the system electronics. These enclosures were placed in harsh environments, and after a number of years in the field experienced unexpected electronic failures in northern climates. Upon closer investigation, it was found that standing water was present at the bottom of the equipment manhole; the exterior of the enclosures had rust present; and rust-colored water was present within the enclosure. During the installation process, the boxes were placed in a manhole near the monitoring sight. The boxes were scratched or dented, which exposed the metal surfaces. Additionally, the salt spread on the roads in winter eventually came into contact with the enclosures as the snow melted. Over time, this exposed surface metal eventually rusted.

The water level inside the manhole could not be controlled, so it was decided that an alternative material should be used for the enclosure that would not rust and could be sealed. A reinforced compression molded electrical enclosure was selected, which offered a variety of sealing options for the door and door configurations.

References

1. Trelleborg AB, "Manufacturing processes," http://www.trelleborg.com (accessed July 30, 2008).
2. Goodyear, "History Overview," http://www.goodyear.com (accessed October 31, 2008).
3. Goodrich, "Goodrich History," http://www.goodrich.com (accessed October 2, 2008).
4. The Open University, "Thermosetting Plastic," Wikipedia, http://en.wikipedia.org/wiki/Thermoset (accessed October 24, 2008).

5. Jeff Butcher, "Compression Molding," BSU, http://www.bsu.edu (accessed October 2, 2008).

Appendix A: Plastics Terms, Definitions, and Examples from A to Z

The plastics industry can be confusing and intimidating because of the different conversion process options and terminology. This appendix is intended to be one of the most extensive lists of terms and definitions available today. Where appropriate, examples have been provided to further clarify a term. Within the definitions, words in italics are found elsewhere in this appendix.

A

Abrasion resistance – (Material): The ability of a material or *mold* to resist scratches and wear.

Abrasive – (Material): Used to describe a *resin* that is harsh on the *finish* of the tooling. For example, glass-filled materials are very abrasive and will change the surface *finish* of the *mold* and increase the *gate* size over an extended period of time.

ABS – (Acronym): Acrylonitrile butadiene styrene.

Acrylonitrile butadiene styrene – (Material): An *engineering* grade *amorphous* material with a middle-range heat deflection temperature and excellent impact properties. This material is chosen for applications like tool handles, electronic housings, cell phones, and other parts requiring high impact. This is a *thermoplastic* ter-polymer combining acrylic for toughness, butadiene for impact, and styrene for gloss.

Additive – (Material): A substance compounded into the *resin* to impart specific properties of the *resin*. It may also diminish or enhance other properties in general. For example, adding *talc* to *polypropylene* gives it better *dimensional stability* and lower cost, but reduces the *impact strength*.

Air trap – (Tooling): Condition caused by a poorly vented *tool* or when two *flow fronts* converge. The air traps will be visible by a *burn* mark or incomplete part feature.

Ambient temperature – (Material): Manufacturing plant or room temperature.

Amorphous polymers – (Material): Materials with no ordered molecular chains. These materials are typically *transparent* and exhibit a broad range of *melting temperatures*. An excellent example is the *commodity resin* styrene.

Anisotropic shrinkage – (Material): Dimensional *shrinkage* that varies from the cross-flow to the down-flow direction of the melted *polymer*. Glass-filled materials will shrink more in the cross-flow direction.

Antioxidant – (Additive): Used to minimize the effects of oxygen-related deterioration of *plastic* parts by preventing the breakdown of molecular chains.

Antistatic – (Additive): Compound added to a *resin* to reduce a static charge on the surface of the part. Examples would be *carbon black* or carbon-impregnated fibers.

A-plate – (Tooling): Metal plate where half of the *cavity* detail and *gate* are typically located.

A-side – (Tooling): The half of the mold assembly that is attached to the *stationary platen* or *extruder* side of the *injection molding* machine. It typically contains half of the part detail. During the injection *cycle*, the *B-side* clamps against the A-side. Also referred to as the *cavity side*, *hot half*, or *stationary side*.

Aspect ratio – (Part design): Ratio of length to width, typically used when referring to a mineral- or glass-filled *resin*.

Assembly – (Part design): (1) Joining two or more parts together. (2) The final design or molded product.

ASTM – (Acronym): American Society for Testing and Materials. Describes sample feature size and tests for physical, mechanical, electrical, flammable, and thermal properties. It is important to note that test plaques do not represent typical wall thicknesses in molded parts.

B

Back pressure – (Process/equipment): Resistance to the flow of oil into the pump reservoir on hydraulic presses as the *screw* rotates backwards. Typically, it is between 50 and 200 psi (0.34–1.38 MPa). A higher back pressure causes more mechanical energy to be transferred to the *pellets* during *plastication*.

Backflow – (Process): During the molding process, when the *injection pressure* is transferred from *pack* to *hold*, a small amount of material will flow back into the *barrel* until the pressure in the *cavity* and the *barrel* are equal or the *gate freezes off*.

Baffle – (Tooling): A blade used to direct the water in a dead-end channel. The blade divides the water so that it flows up one half of the cylindrical channel and down the other side to remove heat from the *mold*.

Balanced runner – (Tooling): *Runner* configuration in which material is delivered to all parts with the same rate of flow. Parts must be identical, not a *family mold*.

Barrel – (Equipment): Contains the *screw, check ring*, and *heater bands*, which together melt and deliver *plastic* to the *mold*.

Birefringence – (Material): Occurs when isotropic materials are deformed such that isotropy is lost in one direction. The *plastic* chains are frozen in the direction of flow.

Blemish – (Molding defect): An imperfection on the surface of the *molded* part.

Blend – (Material): Two or more distinct materials, such as *high impact polystyrene (HIPS). HIPS* is a multiple-phase blend of styrene and rubber. Each one remains discreet, meaning there is no chemical bonding.

Blow molding – (Conversion process): A technique for making hollow parts by injecting air into an extruded *parison* of *plastic* that has been pinched between two *mold* halves.

Blowing agent – (Additive): Used to prevent *sink* when *molding* parts with wall thicknesses that are 0.157″ (4 mm) or more. The blowing agent forms small internal air *voids* or bubbles as the part cools.

Boss – (Part design): Feature which protrudes from the surface of a part. Used to align parts or aid in attaching parts with fasteners or a *secondary operation*.

Bottom clamp plate – (Tooling): Provides the surface for attachment to the *moving side* or *B-side* of the *tool*.

B-plate – (Tooling): Metal plate where the *core* or inner details are typically located.

Breaker plate – (Equipment): Provides support for the *screen pack* and a mounting surface for the *extrusion die*.

Brittle – (Molding defect): Mechanical property condition which describes a part with very low *impact strength*.

B-side – (Tooling): The half of the mold that is attached to the *moving platen* of the *injection molding* machine. During the injection *cycle* the B-side clamps against the *A-side*. Also referred to as the *cold half, core side*, or *moving side*.

Burn – (Molding defect): Occurs when air is trapped in the *mold* and cannot escape during injection. It superheats and chars the *flow front*. A second source of burnt material comes from *degradation* due to excessively long *dwell time* or elevated temperatures in the *delivery system*.

C

CAD – (Acronym): Computer aided design.

CAM – (Acronym): Computer aided machining.

Captive molder – (Manufacturing): Supplier who provides parts or assemblies to a single customer.

Carbon black – (Additive): A powder that is compounded into *resins* to enhance properties like ultraviolet resistance or static dissipation.

Carbon fibers – (Additive): A conductive reinforcement compounded into *resins* to increase the material strength.

Cavity – (Tooling): (1) Typically forms the outside details of the molded part. (2) Another term for the *A-side*, *hot half*, or *stationary side* of the *mold*.

Cavity blocks – (Tooling): Machined to carry the detail of the part. Also referred to as an *insert*.

Cavity identification – (Tooling): A mark, typically a number, which is located on an inconspicuous area of the molded part. In most cases it is located on the underside of the part. This is used to identify cavities of the *mold*, should a quality related issue arise.

Cavity side – (Tooling): The side of the *mold* that forms the exposed or outside surface of the part. Usually referred to as the *A-side*, *hot half*, or *stationary side* of the *mold*.

Check ring – (Equipment): Part of the *screw* tip assembly that prevents *backflow* of *plastic* into the *screw* flights during injection. During *plastication* the check ring moves forward to allow melted *plastic* into the *barrel* in front of the tip.

Chemical resistance – (Material): The ability of a material to withstand chemical attack or *degradation*.

Clamp force – (Equipment): The force, in tons, required to hold the *mold* halves together during the molding *cycle*. Typically it is 3.5 to 4.0 tons per square inch of *projected area*. It can also be referred to as clamping force or clamping pressure.

CNC – (Acronym): Computer numerical control.

Co-extrusion – (Conversion process): Process of extruding a first material followed by a second extruded material, which covers a proportion or all of the first material.

Cold half – (Tooling): The *B-side* or *moving side* of the *mold*.

Cold runner – (Tooling): An unheated *polymer* delivery path to the part which is ejected after each *cycle*.

Cold slug – (Molding defect): A solidified piece of *plastic* from the previous shot which is injected into the next *cycle*. It can appear as a visual defect in line with the *gate*.

Colorant – (Additive): Concentrated colored *pellets* which are compounded with *natural resins* in a specific *letdown ratio* or *concentrate*.

Commodity resin – (Material): A class of *resins* that do not have high temperature properties or high mechanical strength. *Resins* commonly included in this class are *polyethylene* (PE), *polypropylene* (PP), and *polystyrene* (PS). These materials are the most widely used and present the lowest cost per pound.

Compounding – (Material): The combination of *additives* and natural *polymers*.

Compression molding – (Conversion process): A low pressure process used to *mold thermoset* materials.

Compression zone – (Equipment): One of the three *zones* of the *screw* where *resin pellets* are forced against the rotating *screw* flights, thereby providing the mechanical energy to melt the *pellets*. Sometimes referred to as the *transition zone*.

Computer numerical control – (Equipment): Computer controlled machining equipment used in the fabrication of molds.

Concentrate – (Additive): See *colorant*.

Cooling line – (Tooling): Primary means of removing heat from the *mold* to solidify the injected *plastic*. Also referred to as a *water line*.

Cooling time – (Process): After injection, the amount of time, in seconds, to solidify the part for ejection.

Copolymer – (Material): A *polymer* of two or more different *monomers* to impart specific properties.

Core – (Tooling): The piece of the *mold* that forms the internal detail or structure of a part.

Core side – (Tooling): Typically forms the internal details of the *molded* part and is commonly referred to as the *B-side, force side,* or the *moving side* of the *mold*.

Corner identification – (Tooling): Used by tooling makers to determine the orientation of the individual plates of a *mold*. Typically, designated with a metal punched "0" character.

Corona treatment – (Process): A process which is used to prepare a low energy surface, such as *polyethylene* (PE) or *polypropylene* (PP) for a *secondary operation* like *pad printing*.

Crazing – (Molding defect): A very fine crack on the surface of a *molded* part.

Creep – (Part design): Under a long-term load, the *plastic* will permanently deform or creep to relieve the applied stress. An example of this is a *snap fit*, which has members under constant stress to achieve the locking or joining of two parts.

Critical dimension – (Part design): Feature of the part which is essential to the function of the part or product.

Cross-linked high-density polyethylene – (Material): A *commodity*-grade *semi-crystalline* material with a low heat deflection temperature, good *chemical resistance*, and good *elongation* properties. This material is used in tubing and medical applications.

Cross-linked polyethylene – (Material): A *commodity*-grade *semi-crystalline* material with a low heat deflection temperature, good *chemical resistance*, and good *elongation* properties. This material is used in radiant heating systems, chemical and oil transportation containers, and plumbing. Commonly referred to as *PEX*.

Crush – (Tooling): An allowance built into the tooling to provide a positive seal between the *mold* halves.

Crystalline – (Material): Materials with an ordered structure. They exhibit a discreet melting point and are typically non-*transparent* with good to excellent *chemical resistance*. A common example is *polypropylene* (PP).

Custom molder – (Manufacturing): Supplier who provides parts or assemblies to customers at large.

Cycle – (Process): All steps required to *mold* an individual part. This would include mold closed, inject, cooling, mold open, and part ejection.

Cycle time – (Process): Total time, in seconds, required to *mold* one part or a set of parts in a *tool*.

Cycling to the runner – (Tooling): Condition in which a part cannot be ejected until the *runner* has cooled. This typically occurs when molding small or thin wall section parts with a larger *runner*.

D

Daylight – (Equipment): Maximum space between *tie bars*. This defines the maximum *mold* size that can be mounted into the press.

Degradation – (Process): Loss of material properties due to overheating or repeated processing.

Delivery system – (Equipment): The path of the melted *resin* through the *barrel* and the *manifold* or *runner* up to the *gate*.

Die – (Tooling): Primary plate used in the *extrusion* process to create films and profiles.

Differential cooling – (Process): When nonuniform temperatures are applied to the *cooling lines* of a mold. For example, the *A-side* may be set at 120°F (49°C), while the *B-side* is held at 80°F (27°C). This cooling practice is occasionally used to bias the part to minimize *warp*. This will increase *tool* wear and may not be a good practice.

Dimensional stability – (Part design): Characteristic of a part to retain its shape after cooling.

Direct gate – (Tooling): *Tool* design in which the *resin* enters directly into the part from the *sprue bushing*. This is a runnerless design intended for very large parts.

Down stops – (Tooling): Component of an *injection mold* used to limit the travel of the *ejector plate*. It defines the home position for the ejector assembly.

Draft – (Tooling): Allowance needed for removal of the part from the *mold* without damage. For *injection molding* the typical draft is 1° which is 0.017″ per inch or 0.43 mm per mm.

Drag marks – (Molding defect): Condition on the surface of the part which occurs when inadequate *draft* is present. This may be seen in the form of lines in the direction of ejection.

Drawdown – (Process): Used in the *extrusion* process, it is the ratio of the thickness of the *die* opening to the thickness of the desired part.

Drool – (Process): Unintended flow of *plastic* through a *nozzle, sprue,* or part *gate.*

Drying – (Process): Thermal process used to remove moisture from *hygroscopic* materials before processing.

Ductility – (Material): The ability of a material to deform without fracturing.

Durometer – (Material): A *hardness* rating system for thermoelastic materials. This system uses the *Shore A* scale.

Dusting the parting line – (Tooling): A grinding adjustment made to a *mold* that has excessive *parting line, flash,* or mismatch.

Dwell time – (Process): Used in ultrasonic welding and pad bonding. The amount of time that the horn (ultrasonic welding) or the pad (*pad printing*) is in contact with the part.

E

EBM – (Acronym): Extrusion blow molding.

Edge gate – (Tooling): A *gate* design where *plastic* is injected into an edge of the part. This *gate* requires removal after molding by mechanical means.

EDM – (Acronym): Electrical discharge machining.

Ejector blade – (Tooling): A rectangular component used to assist removal of the part from the *mold.*

Ejector pin – (Tooling): A round component used to assist removal of the part from the *mold.*

Ejector pin marks – (Part design): Visual marks on the interior of the part left from the heads of the *ejector pins.*

Ejector plate – (Tooling): Attached to the *ejector retainer plate.* It registers the *ejector pins* to the surface of the *cavity.*

Ejector retainer plate – (Tooling): Holds the *ejector pins* or *ejector blades* in the correct position. It is attached to the *ejector plate.*

Ejector sleeve – (Tooling): A cylindrical component used to assist ejection of a screw *boss.*

Ejector system – (Tooling): All plates and components that facilitate part removal upon mold open. Specifically it includes *ejector plate, ejector retainer plate, ejector pins, down stops,* and *return pins.*

Elastomer – (Material): Flexible *thermoplastic.* Common application is overmolded tool grips.

Electric injection molding machine – (Equipment): *Injection molding* press that uses electric power to clamp the *mold* and inject *plastic*.

Electrical discharge machining – (Equipment): The process where a graphite electrode is machined to the geomery of the part to be molded. A current is passed through the electrode causing a spark. The spark causes the metal to melt. The process is repeated until the complete geometry is burnt into the mold.

Elongation – (Material): The measure of a material's *ductility* or ability to be stretched.

Engineering resins – (Material): A class of *resins* that have higher heat stability and *impact strength*. For example: *polyamide* (PA), *polyetheretherketone* (PEEK), *polyetherimide* (PEI), *polyethylene terephthalate* (PET), *polycarbonate* (PC), and *styrene acrylonitrile* (SAN).

Extruder – (Equipment): A machine designed to melt or plasticize *pellets*. It includes a tubular *barrel, heater bands*, a rotating *screw*, and a *hopper*. The extruder unit is used with *injection molding, extrusion,* and *blow molding* equipment.

Extrusion – (Conversion process): The continuous process whereby *plastic pellets* are melted in a heated *barrel* and forced through a shaping plate or *die*.

Extrusion blow molding – (Conversion process): A popular process used to produce containers, the *extruder* forms a *parison* that is captured between two *mold* halves and inflated to form the final shape.

F

Family mold – (Tooling): A *mold* that contains different part configurations of the same product. For example, both halves of a tape dispenser.

Fan gate – (Tooling): A triangular shaped edge *gate* used for thick sectioned parts. Advantages: low stress, slow *fill* without freezing, and even flow.

Feed zone – (Equipment): The *screw* has three *zones*. The feed zone has *screw* flights to accept discrete *pellets* from the *hopper*. It moves the *pellets* to the *transition zone*.

Fill – (Process): (1) The first stage of the injection process. The *cavity* volume is approximately 95%–98% complete. This is done at high injection pressure for a short period of time. (2) Injecting of *resin* into a *mold*.

Fill pressure – (Process): Machine pressure (psi) during the *fill* part of the *cycle*.

Fill time – (Process): (1) Amount of time required to *fill* the *mold* approximately 95%–98%. (2) The first stage of the injection *cycle*.

Filler – (Additive): A compound or material blended with *virgin resin* to enhance properties or reduce cost.

Finish (Surface) – (Tooling): The visual appearance or *texture* of a surface.

Flame retardant – (Additive): A material blended with the *resin* to impart fire-resistant properties.

Flash – (Molding defect): Excess material on the parting line, ejector pins, or inserts which can be the result of excessive molding pressure, *tool* wear, inadequate *clamp force*, or a poor fit between *mold* components.

Flash gate – (Tooling): Thin area across the *parting line* where *plastic* is injected into the *cavity*. It is used for long, flat, thin walled parts. *Gate* size is 0.010″–0.020″ (0.254–0.508 mm) thick; the land length is 0.020″–0.040″ (0.508–1.016 mm) long.

Flexural modulus – (Material): Resistance of a material to bending under a load.

Flight depth – (Equipment): The root dimension of the *extruder screw* between two adjacent flights.

Flow front – (Process): The leading edge of the *polymer* as it advances during the injection *cycle*.

Flow lines – (Part design): A visual mark on the surface of the part where two *flow fronts* meet. Flow lines can be very visible when molding metallic colors.

Foaming agent – (Additive): A chemical compound used in the *structural foam molding* process. It forms air *voids* in the wall sections to minimize *sink*.

Freeze off – (Process): As the injected *resin* cools, it solidifies and prevents additional material from entering the *cavities*.

G

Gate – (Tooling): The channel through which the *resin* enters the part being *molded*.

Gate location – (Tooling): The position on the *tool* or part where the material enters the *cavity*.

Gaylord – (Material): A large square corrugated box with a *plastic* liner which typically measures 4′ × 4′ × 4′ (1.22 × 1.22 × 1.22 m) and holds approximately 1000 pounds (453.6 kg) of materials. The box also has a cover that can be removed.

Glass fibers – (Additive): A reinforcement material blended to impart stiffness.

Glass transition temperature – (Material): The softening temperature of *amorphous resins*. Abbreviated as T_g.

Go/no-go gauge – (Process): A quality acceptance tool.

Guided ejection – (Tooling): The *ejector plate* assembly is aligned during travel with precision pins and bushings.

Gusset – (Part design): Internal supporting wall.

H

Hardness – (Tooling): A measurement of durability. *Tool* materials are measured using the *Rockwell scale* and *elastomers* are measured using the *Shore A* scale.

HDPE – (Acronym): High density polyethylene.

Heat treating – (Tooling): The process followed to harden a steel *mold*.

Heater band – (Equipment): A component of the *extruder*. It maintains the *polymer* melt at a constant temperature for molding.

High-density polyethylene – (Material): A *commodity*-grade *semi-crystalline* material with a low heat deflection temperature and excellent solvent resistance properties. This material is used as a low cost general-purpose material in many conversion processes.

High impact polystyrene – (Material): A *commodity*-grade *amorphous* material with a middle heat deflection temperature and excellent impact properties. This material is used in tool housings, handles, and consumer electronic remotes.

High-molecular-weight high-density polyethylene – (Material): A *commodity*-grade *semi-crystalline* with a low heat deflection temperature, excellent *impact strength*, and *chemical resistance* properties. This material is used for joint and bone implants and medical devices.

HIPS – (Acronym): High impact polystyrene.

HMW-HDPE–(Acronym): High-molecular-weight, high-density polyethylene.

Hobbing – (Tooling): Occurs when excess *flash* or material from the previous shot is present when the *mold* closes. The excess material will deform or make an impression in the *mold* plates.

Hold – (Process): During the molding cycle, low pressure is applied to the *cavity* or *cavities* to prevent *backflow* into the *nozzle* and to allow the *gate* to freeze.

Homogenous – (Material): A material for which local variations in composition are negligible.

Homopolymer – (Material): A *polymer* resulting from the polymerization of a single *monomer*. For example, *polyethylene* (PE).

Hopper – (Equipment): Piece of auxiliary equipment which dries and gravimetrically delivers *resin* to the throat of a barrel *extruder*.

Hot half – (Tooling): See *A-side*.

Hot manifold – (Tooling): Runnerless method of delivering *plastic* to the *mold*.

Hot runner – (Tooling): Heated *runner* that maintains temperatures and does not allow the material to solidify.

Hot side – (Tooling): See *A-side*.

Hot tip – (Tooling): Located in the *mold* and acts like a *gate* into the part. It is used with a runnerless system and is cycled on and off to allow material to flow into the *cavity*.

Hydraulic injection molding machine – (Equipment): *Injection molding* press that uses hydraulic power to clamp the *mold* and inject *plastic*.

Hydrophilic – (Material): Material that easily absorbs moisture.

Hydrophobic – (Material): Material that does not easily absorb moisture.

Hygroscopic – (Material): The tendency of a material to absorb moisture.

I

IBM – (Acronym): Injection blow molding.

Impact modifier – (Additive): A material compounded to enhance the impact resistance properties of a material. For example, butadiene is blended with styrene to create high impact styrene.

Impact strength – (Material): The ability of a material to withstand direct impact. A typical example is a 20 oz (591 mL) soda bottle.

Injection blow molding – (Conversion process): Hybrid process which combines *injection molding* in the first step and *blow molding* in the second step.

Injection mold – (Tooling): Specific to injection molding, a system of plates, which include the part detail, means of cooling, and part ejection. Also see *mold*.

Injection molding – (Conversion process): Process where melted *resin* is injected under pressure into a *mold*.

Injection pressure – (Process): Force required to *fill* and *pack* a molded part.

In-mold decorating – (Process): Integrating or placing a label directly into the *cavity* and molding the part onto it.

Insert – (Tooling): Removable section of the *mold* that contains part detail. This method is often used when the part contains small features or difficult-to-*polish* details.

Insert molding – (Process): Procedure which requires the placement of an object into the *mold* prior to the shot. A common example is the placement of a threaded *insert*.

Isotropic shrinkage – (Material): The same post-molding *shrinkage* in both the cross flow direction and down flow direction of the part. This is specified in the *material data sheets* in units of inch per inch.

Izod impact test – (Material): A test method for comparing impact properties of materials.

J

Jetting – (Molding defect): Visual defect that originates at the *gate* and appears in a serpentine pattern in the molded part.

Jig – (Equipment): A fixture for *assembly* or testing.

K

Knife edge – (Tooling): A thin steel condition in the *mold* caused by improper part design.

Knit line – (Part design): Condition that exists when two or more material *flow fronts* merge together. An example is material flowing around a hole formed by a pin or material delivered from multiple *gates*.

Knockout holes – (Tooling): Holes in the *bottom clamp plate* which allow the *knockout pins* of the press to move the *ejector plate* forward as the *mold* opens.

Knockout pin – (Tooling): Bars attached to the press which are used to move the *ejector plate* forward.

L

Laminar flow – (Process): *Reynolds number* between 2000 and 4000.

LDPE – (Acronym): Low-density polyethylene.

Leader pins – (Tooling): Steel pins that guide the *mold* halves together as the *mold* opens and closes.

Letdown ratio – (Material): Material manufacturer's specification of a color *concentrate* to a *natural resin*.

Linear low-density polyethylene – (Material): A *commodity*-grade *semi-crystalline* material with a low heat deflection temperature, high *tensile strength*, and impact properties. This material is used in toys, buckets, and pipes.

LLDPE – (Acronym): Linear low-density polyethylene.

Locator ring – (Tooling): Used to align the *mold* to the center of the platen. This is important so the pressure is distributed evenly among the *tie bars*.

Low-density polyethylene – (Material): A *commodity*-grade *semi-crystalline* material with a low heat deflection temperature, good *tensile strength*, good *chemical resistance*, and good impact properties. This material is used as a low cost general-purpose material in containers, bottles, bags, and tubing.

Lubricant – (Additive): Material compound added to aid in the release or friction bearing properties of a material. Two common *additives* are silicone and mineral oil.

M

Manifold – (Tooling): Multiple path *melt delivery system.*

Material data sheet – (Material): Manufacturer's specifications based on standard testing properties that include general, mechanical, thermo, electrical, and *flame retardant* values.

Melt delivery system – (Equipment): The path *resin* flows through the *mold*. This can be a *cold runner* or a *hot manifold/runner* design.

Melt flow rate – (Material): Test method that measures the amount of material that flows through standardized test equipment in 10 minutes. Results are recorded in grams per 10 minutes. Low melt flow values tend to be used for *blow molding* or extruding, while high melt flow values are used for *injection molding.*

Melt index – (Material): See *melt flow rate.*

Melt transition temperature – (Material): Temperature at which the *crystalline* materials can be extruded or molded. Abbreviated as T_m.

Melting temperature – (Material): Temperature at which a material is completely in a liquid state.

Metering zone – (Equipment): Last *zone* of the *screw*. The *flight depth* is very shallow. The purpose of the metering zone is to feed material forward through the *check ring* into the *barrel*.

Mold – (Tooling): (1) Used in various conversion processes to describe the system of plates or halves that contain the part detail, means of cooling, and part ejection. (2) The action of creating *plastic* parts using a rigid form. For example, "the parts will be molded next week."

Mold base – (Tooling): Un-machined set of all plates and basic components that are then machined to produce a part.

Mold release – (Process): Topical application on the *mold* of a spray or wax to assist in part ejection.

Mold shrinkage – (Tooling): Machining allowance applied to the *mold* to compensate for post molding material *shrinkage. Shrinkage allowance* can be obtained from the *material data sheet.*

Mold temperature – (Process): Temperature at which the *mold* is maintained to complete the process.

Molded in stress – (Process): Occurs because molecular chains are stretched as they flow under high pressure into the *cavity*. Areas of concentration occur at sharp transitions and near the *gate*. Stress can also be effected by very rapid cooling.

Molecular weight – (Material): Total atomic weight of all atoms contained in a *molecule*.

Molecule – (Material): The smallest particle of a substance.

Monomer – (Material): A *molecule* that can be combined with itself or different *molecules* to create a *polymer*.

Moving platen – (Equipment): The plate on the *injection molding* press on which the *B-side* of the *mold* is mounted.

Moving side – (Equipment): The *B-side* of the *injection molding* press.

Multi-cavity mold – (Tooling): A *mold* that contains more than a *single cavity* of the same part.

Multi-shot injection molding – (Conversion process): A process that utilizes a press with multiple *barrels* to produce a part that consists of multiple materials.

N

Natural resin – (Material): Base *resin* without any *additives, colorants*, or *fillers*.

Nozzle – (Equipment): Located at the end of the *barrel extruder* and functions as the interface to the *mold* through the *sprue bushing*.

NPE – (Acronym): National Plastics Exposition.

O

Offset gate – (Tooling): The location of a *gate* into a part that results in an unbalanced flow of material into the *cavity*. This is typically done to change the location of a *knit line* within the part or to align the flow of material in a specific location.

Olefin – (Material): Family of resins that includes *polyethylene* (PE), *high-density polyethylene* (HDPE), *linear low-density polyethylene* (LLDPE), *ultra high molecular polyethylene* (UHMPE), *ultra low-density polyethylene* (ULDPE), and *polypropylene* (PP).

Opaque – (Material): Characteristic of a material that blocks the transmission of light through the material.

Overmold – (Process): A process by which a second material is molded onto a first material.

P

PA – (Acronym): Polyamide.

Pack – (Process): Second stage of the *injection molding cycle*.

Pad printing – (Secondary operation): A method for decoration or marking a part whereby a silicone pad transfers the image from an etched plate to the part. This is an ideal process for marking contoured parts.

Parison – (Process): A hollow tube of heated *plastic* that is used in the *blow molding* process. The tube is surrounded by the *mold* and compressed air is forced through the inside of the hollow tube to expand it within the *mold*.

Parting line – (Tooling): The natural division between the *cavity* and *core side* of the part. This corresponds to the *A-side* and *B-side* of the *mold*. It may be a visible detail or line, typically less than 0.005″ (0.127 mm) high and wide present in the molded part.

PBT – (Acronym): Polybutylene terephthalate.

PC – (Acronym): Polycarbonate.

PE – (Acronym): Polyethylene.

PEEK – (Acronym): Polyetheretherketone.

PEI – (Acronym): Polyetherimide.

Pellets – (Material): Granular form of a *resin*.

PET or PETE – (Acronym): Polyethylene terephthalate.

PEX – (Acronym): Cross-linked polyethylene.

Plastic – (Material): A category of materials that can be readily formed into many shapes and objects.

Plastication – (Process): Occurs when individual *resin pellets* are compressed against the *screw* flights as the *screw* is rotating. This creates friction and heat, which melts or plasticates the *polymer*.

Plasticizer – (Additive): Enhances the flow characteristics of materials.

Plating – (Tooling): A surface treatment applied after fabrication to a mold or die in order to obtain certain properties. Common treatment types are wear, durability, enhanced part release, antioxidation, etc.

PM – (Acronym): Preventative maintenance.

Pocket – (Tooling): Machined area that accepts a *cavity block* or *insert*.

Polish – (Tooling): Fine surface *finishes* of a *mold* or *die*. Standard *finishes* are specified by SPI and are defined by the alphanumeric designations A1 (mirror) to C3 (light frost). For example, CD and DVD *molds* have an A1 *finish*. Rough surface *finishes* are referred to as *textures*.

Polyamide – (Material): An *engineering*-grade *semi-crystalline* material with high heat deflection, excellent *chemical resistance*, high modulus, and high *impact strength*; however, it absorbs water. This material is used in high temperature applications.

Polybutylene terephthalate – (Material): An *engineering*-grade *semi-crystalline* material with high heat deflection and high solvent resistances. This material is used in the electronics industry for insulators.

Polycarbonate – (Material): An *engineering*-grade *amorphous* material with a high heat deflection temperature and excellent impact properties. This material is used to mold CDs, DVDs, and safety shields.

Polyetheretherketone – (Material): An *engineering*-grade *semi-crystalline* material that has an extremely high heat deflection temperature and is extremely rigid. This material is used in the electronics industry for solder wafer trays.

Polyetherimide – (Material): An *engineering*-grade *amorphous* material with a very high heat deflection temperature and high *flexural modulus*. This material is used in high heat applications.

Polyethylene – (Material): A *commodity*-grade *semi-crystalline* material with a low heat deflection temperature, good *chemical resistance*, and good impact properties. This material is used as a low cost general-purpose material in many conversion processes.

Polyethylene terephthalate – (Material): A *commodity*-grade *semi-crystalline* material with a very high heat deflection temperature, good impact strength, and good barrier properties. This material is used in the *blow molding* process for beverage containers.

Polymer – (Material): A large *molecule* composed of repeating subunits connected by covalent bonds.

Polypropylene – (Material): A *commodity*-grade *crystalline* material with a low heat deflection temperature and good *chemical resistance* properties. This material is used as a low cost general-purpose material in many conversion processes, in particular containers and medical applications.

Polystyrene – (Material): A *commodity*-grade *amorphous* material with a middle range heat deflection temperature and poor solvent resistance properties. This material is used as a universal material blended with *additives* to enhance the base properties of the *resin*.

Polyurethane – (Material): A class of *polymers* that can be rigid, flexible, or liquid depending upon the chemistry.

Polyvinyl chloride – (Material): A *commodity*-grade *amorphous* material with a low heat deflection temperature and good impact properties. This material is used in the *extrusion* of plumbing pipes, residential siding, and window trim.

PP – (Acronym): Polypropylene.

Preform – (Process): Typically an *injection molded* part which is later reheated to create the final part. An example of this would be a 2-liter soda bottle.

Pressure transducer – (Process): A device for indirectly measuring the pressure at a specific point in the *mold*.

Preventative maintenance – (Tooling): The practice of scheduled inspection and repair of *dies, molds,* and equipment.

Primary runner – (Tooling): The material path from the *sprue* to the *secondary runner.*

Process monitoring system – (Process): Computer based hardware/software interface to the *mold* and the press. This equipment is specifically used in *injection molding.*

Profile extrusion – (Process): Specific to non-tubular *extruders* or *extrusions.*

Projected area – (Tooling) : Area in square inches of the part on the *A-side* of the *mold.* The projected area is used in the calculation of the required *clamp force.*

Prototype tool (mold) – (Tooling): *Single cavity mold* that is used to produce first generation parts.

Pry slot – (Tooling): Located on one corner of an *injection mold* and used to assist in the separation of *mold* plates.

PS – (Acronym): Polystyrene.

PU – (Acronym): Polyurethane.

Purge – (Process): Process of using one material to clear the *barrel* of the previous material.

PVC – (Acronym): Polyvinyl chloride.

Q

Quench – (Process): Rapid cooling of a *crystalline* or *semi-crystalline* material.

R

Race tracking – (Part design): Occurs in *injection molded* parts when material flows around the outer edge of the part instead of filling from the *gate* outward.

Reaction injection molding – (Conversion process): This process is similar to *injection molding* except it uses a two-part liquid *polymer* system which reacts in the *mold* to form a *thermoset* part.

Reciprocating screw – (Equipment): Flighted shaft that on the reverse stroke plasticizes *pellets* and then on the forward stroke acts as a ram to inject the *plastic* into the *mold.*

Recovery time – (Process): Length of time that the *screw* rotates to plasticize material.

Recycle symbol – (Tooling): A marking on parts designating *plastic* waste stream.

Regrind – (Material): *Resin* that has completed one or more heat cycles and has been ground into smaller size pieces so it can be put into the *hopper* for another molding *cycle*.

Reinforced plastic – (Additive): *Resin* that contains glass or mineral fibers which increase the *flexural modulus* and impact properties.

Release agent – (Additive): *Resin additive* that facilities part removal.

Residence time – (Process): Maximum amount of time a single shot remains in the *barrel* during injection.

Resin – (Material): *Plastic polymer* or material.

Return pins – (Tooling): Fail safe feature in the *mold* to ensure that the *ejector pins* retract prior to the *mold* closing.

Return springs – (Tooling): Located on the *return pins* to assist in the retraction of the *ejector pins* prior to the *mold* closing.

Reynolds number – (Process): A number which indicates laminar, transient, or *turbulent flow*. Numbers less than 2300 are considered laminar in flow; numbers between 2300 and 4000 are considered transient in flow; and numbers greater than 4000 are turbulent in flow. *Cooling lines* in the *mold* should have *turbulent flow*.

Rheology – (Process): The study of material flow.

Rib – (Part design): A projection from the interior surface of the molded part. These are typically used to increase the strength of a part or remove material from a large wall section.

Ring gate – (Tooling): A gating technique used to balance round parts.

Rockwell scale – (Tooling): A system for measuring the *hardness* of steel.

Runner – (Tooling): The material path to the *cavities*. Also see *cold runner* and *hot runner*.

Runner shut-off – (Tooling): A mechanical *shut-off* which blocks or directs the flow of material to one or more *cavities*. Also referred to as a *shut-off*.

S

SAN – (Acronym): Styrene acrylonitrile.

SBM – (Acronym): Stretch blow molding.

Scientific molding – (Process): Method that standardizes the molding *cycle* based upon velocity and transfer pressure. For any one *mold*, the intent is to use one process in multiple machines and get the same results.

Scrap – (Material): Extra material generated during the molding process such as *runners* and parts that are not to specifications. For *thermoplastics*, the scrap typically becomes *regrind*.

Screen pack – (Equipment): A series of wire screens with varying mesh sizes used to filter out possible contaminants or unmelted *resins* before they reach the *die*.

Screw – (Equipment): Short for *reciprocating screw*.

Screw speed – (Process): Revolutions per minute applied to the *screw* during recovery. Typical speeds are 50–200 RPMs and higher RPMs impart more energy to melt the *resin* as well as mix color *concentrates*.

Screw travel – (Process): Distance in inches the *screw* travels forward during injection.

Secondary operation – (Process): Process that occurs to a part after it is molded. For example, two parts can be welded together using ultrasonic welding, or labeling can be applied using *pad printing*.

Secondary runner – (Tooling): The material path from the *primary runner* to the individual *cavities*.

Semi-crystalline – (Material): *Polymers* that have ordered regions of chain alignment and area of random molecular chain alignment. Common examples are *polyethylene* (PE) and *polyamide* (PA). These materials will shrink more than *amorphous* materials.

Shear rate – (Process): A measure, in reciprocal seconds, of the velocity gradient or the rate that the shear is applied.

Sheet extrusion – (Conversion process): Materials created through a flat *die* instead of a profile *die* or *injection mold*.

SHIPS – (Acronym): Super high impact polystyrene.

Shore A – (Material): A *hardness* scale for soft materials.

Short shot – (Process): Produced when a part has not been completely filled during a molding *cycle*.

Shot size – (Process): Weight in grams or ounces of the material delivered during injection. Also referred to as a shot.

Shrinkage – (Material): The amount that a molded part contracts while it is cooling. This value is measured in inch per inch and is listed on *material data sheets*.

Shrinkage allowance – (Tooling): Factor used in the calculation of *mold* feature dimensions. This is expressed in inch per inch units.

Shut-off – (Tooling): See *runner shut-off*.

Side lock – (Tooling): Tooling accessory mounted directly to the *mold* used to align the *A-side* and *B-side* of the *injection mold* during the molding *cycle*.

Single cavity mold – (Tooling): A *mold* that contains one *cavity* of the part.

Sink – (Molding defect): (1) Post-molding appearance defect caused by the intersection of a thick and thin feature. (2) A slight depression in the surface of a molded part.

Slide – (Tooling): Creates part features perpendicular to the *parting line* of an *injection mold*.

Snap fit – (Part design): Mechanical joining of parts typically using a combination of molded features like a cantilevered beam and hook that mates with an opposing feature.

SPE – (Acronym): Society of Plastics Engineers.

Splay – (Molding defect): Visual lines on the surface of an *injection molded* part. Excess moisture (improper *drying*) is the cause of this molding defect.

Sprue – (Tooling): Material path from the *nozzle* to the *runner*.

Sprue bushing – (Tooling): A machined *mold* component which provides an interface between the *mold* and *nozzle*.

Sprue gate – (Tooling): A runnerless *tool* design.

Stabilizer – (Additive): Chemical *additive* to prevent deterioration of material properties. Examples include UV and antioxidants.

Stack molds – (Tooling): A special *mold* design with two *parting lines*.

Stationary platen – (Equipment): The nonmovable side of an *injection molding* press where the *extruder* and *hopper* are located.

Stationary side – (Equipment): The *A-side* of the *injection molding* press.

Stress concentration – (Part design): Sharp feature transitions. An example of this is a lack of a radius at the base of the wall.

Stress crack – (Molding defect): Material fracture caused by improper design or excessive *injection pressure*.

Stretch blow molding – (Conversion process): A two-step process for forming containers.

Stripper plate – (Tooling): Used to aid in the ejection of a part when *ejector pins* are not effective.

Structural foam molding – (Conversion process): Process which uses *resins* with *blowing agents* to produce parts with wall sections that are greater than 0.187″ (4.75 mm).

Styrene acrylonitrile – (Material): A *engineering*-grade *amorphous* material with a middle range heat deflection temperature but brittle. Typically this material is used in toys, viewing windows, and drinking water applications.

Sub gate – (Tooling): See *tunnel gate*.

Suck back – (Process): Technique used to break the *sprue* from the *nozzle*.

Super high impact polystyrene – (Material): A *commodity*-grade *amorphous* material with a middle range heat deflection temperature and superior impact properties. Typically this material is used in hand tools, industrial handheld devices, and other applications where impact is a concern.

Support pillar – (Tooling): A post attached to the *bottom clamp plate* of an *injection mold* that passes through the *ejector plate* assembly and makes contact with the *support plate*. It prevents deflection of the *B-plate* while under injection pressure during the molding *cycle*.

Support plate – (Tooling): Plate that is located between the *B-plate* and the side rails of the *bottom clamp plate* of an *injection mold*. It functions to support the *B-plate* during injection.

T

Tab gate – (Tooling): A *gate* design which is used to minimize pressure loss through the *gate* and provide fast filling of large parts.

Talc – (Additive): A *filler* or extender.

Tensile bar – (Material): ASTM standard test sample used to measure *elongation* properties.

Tensile strength – (Material): Measure of force required to stretch and break a material.

Tensile test – (Material): ASTM procedure to measure properties.

Texture – (Tooling): Rough surface *finish* on a *mold* or *die* which is created by acid etching or the EDM process. Fine surface *finishes* are referred to as *polishes.*

Thermoforming – (Conversion process): A process that converts sheets (continuous or individually) and films into a formed finished part. Sheets are heated and drawn by vacuum or air pressure.

Thermoplastic – (Material): Group of *plastics* that can be subjected to multiple heating cycles to form parts by various conversion processes like *injection molding, extrusion, blow molding,* and *thermoforming.*

Thermoplastic elastomer – (Material): Group of *copolymers* which consist of a blend of *thermoplastic* and rubber *monomers.* Also referred to as a *thermoplastic* rubber. Often this material is *overmolded* to provide a soft touch feature to products.

Thermoset – (Material): Group of *plastics* that can be heated only once. Parts are formed by an irreversible chemical reaction. *Thermosets* are stronger than *thermoplastics,* can withstand higher heat, and are typically *compression molded.*

Thrust bearing – (Equipment): Used in an *extruder* to prevent the *screw* from moving backward in the *barrel* and absorbs the force generated by the *screw* as it rotates to melt the material.

Tie bar – (Equipment): Allows the *clamping forces* to develop between the moving and stationary halves of the *injection molding* press.

Tie bar spacing – (Equipment): (1) Physical measurement of the minimum distance between press *tie bars.* (2) The maximum mold size a particular *injection molding* press can accommodate.

Tie strap – (Tooling): A flat metal bar bolted across the *parting line* of the *mold* to prevent accidental separation during transportation and handling.

Tonnage – (Equipment): The maximum mechanical *clamping force* of an *injection mold* press.

Tool – (Tooling): See *mold*.

Tooling print – (Tooling): Documentation describing *cavity* detail and plate designs.

Top clamp plate – (Tooling): Provides the surface for attachment to the *stationary platen* or *A-side* of the *tool*. It also contains the *locator ring* and *sprue bushing*.

TPE – (Acronym): Thermoplastic elastomer.

TPR – (Acronym): Thermoplastic rubber.

Trade name – (Material): Name given by a company to identify a resin from a similar resin of another company. For a short list of *plastic* trade names visit the following Web site: http://www.polymerweb.com/_misc/tradenam.html.

Transfer molding – (Process): Form of *compression molding* for a multiple part mold requiring *thermoset* materials. *Resin* is placed on an intermediate plate or *cavity* that has, in turn, individual *sprues* to the *cavity* plates below. The *resin* is heated and compressed through the transfer plate into multiple *cavities*.

Transition zone – (Equipment): The second *screw* zone where *pellets* melt against the *screw* flights.

Translucent – (Material): Characteristic of a material that allows partial transmission of light through the material.

Transparent – (Material): Characteristic of a material that allows a high transmission of light through the material.

Tunnel gate – (Tooling): A *gate* design which allows automatic separation of the part from the *runner* upon ejection. Also referred to as a *subgate*.

Turbulent flow – (Process): *Reynolds number* above 4000. Applies to *cooling lines*.

Twin screw – (Equipment): High capacity *extruder* design.

Two-shot – (Process): A molding method to create parts using two different materials.

U

UHMPE – (Acronym): Ultra high molecular polyethylene.

UL – (Acronym): Underwriters Laboratory Incorporated.

ULDPE – (Acronym): Ultra low-density polyethylene.

Ultra high molecular polyethylene – (Material): A *commodity*-grade *semi-crystalline* material with a low heat deflection temperature and excellent wear resistance properties. This material is used for gears, bearings, and limb implants.

Ultra low-density polyethylene – (Material): A *commodity*-grade *semi-crystalline* material with a low heat deflection temperature, good *chemical resistance*, and good impact properties. This material is used as a low cost general-purpose material in many conversion processes.

Ultraviolet stabilizers – (Additive): Inhibitors blended into materials used in outdoor applications.

Undercut – (Part design): A condition when a *plastic* feature is trapped by steel in the *mold*.

Uniform cooling – (Tooling): *Mold* design which spaces *cooling lines* evenly around the part.

Up stops – (Tooling): Attached to the *support plate* to prevent the ejector assembly from colliding into the *support plate*.

V

Vacuum forming – (Conversion process): See *thermoforming*.

Valve gate – (Tooling): Mechanical gating system (a pin that opens and closes to allow material to flow) that eliminates *runners* and allows for low pressure drop across the *gate*. These systems are used with *hot runner* systems.

Vent – (Tooling): Path for air to escape from a closed mold. Vent depths depend upon the material processed and vary between 0.0005″ and 0.0015″ (0.013 and 0.038 mm). Vents are necessary to prevent burning at the end of *fill* and to allow the part to *fill* completely.

Venting – (Tooling): Allows air to escape from closed mold as *plastic* enters through the *gate*. Typically, vents are spaced 1.5″–2.0″ (3.8–5.1 cm) apart at the *parting line* and end of flow.

Vestige – (Tooling): Excess material from the *gate*.

Virgin material – (Material): Material that has not been processed. Also referred to as virgin resin.

Viscosity – (Process): A measure of resistance to flow. The lower the viscosity, the less pressure required to *fill* the *mold*.

Void – (Molding defect): Internal absence of material.

W

Warp – (Molding defect): Non-flat condition caused by the molding process or marginal part design.

Water line – (Tooling): See *cooling line*.

Weld line – (Part design): Location on a molded part where the two *flow fronts* meet. This will always occur around holes in the part, or when multiple *gates* are utilized to *fill* the part.

Wire EDM – (Tooling): One of the processes used to fabricate *mold* details.

X

XHDPE – (Acronym): Cross-linked high-density polyethylene.

X-ray fluorescence – (Process): Analytical method for analyzing or identifying materials at an elemental level. The spectrum is specific to the *polymer* sampled and the process conditions it experienced.

Y

Yield point – (Material): The point at which a material fails.

Z

Zone – (Equipment): The three sections of a *screw*: *feed zone, transition zone,* and *metering zone*.

Zone profile – (Process): Temperature gradient applied to the *barrel*.

Appendix B: Conversion Process Key Characteristics

The following chart compares the key characteristics of the conversion processes described in this book. Key characteristics include volume, material section, part cost, part geometry, part size, tool cost, cycle time, and labor.

- *Volume* describes the typical quantity of parts molded using a particular process.
- *Material section* provides the extent of material options available for processing.
- *Part cost* compares the relative piece part costs of processes.
- *Part geometry* is a gauge of simple to complex features that can be molded.
- *Part size* is the description of the overall measurements.
- *Tool cost* compares mold cost fabrication.
- *Cycle time* is the time required to mold a complete part.
- *Labor* is the measure of how a process is performed to obtain a finished part.

Conversion Process Key Characteristics

	Chapter 1: Injection Molding	Chapter 2: Extrusion	Chapter 3: Blow Molding	Chapter 4: Thermoforming		Chapter 5: Reaction Injection Molding		Chapter 6: Rotational Molding	Chapter 7: Compression Molding	
				Sheet Stock	Roll Stock	Soft Tool	Hard Tool		Vertical	Transfer
Volume	High	High	High	Low	High	Low	High	Medium	Medium	High
Material selection	Extensive	Moderate	Moderate	Moderate	Moderate	Limited	Limited	Limited	Limited	Limited
Part cost	Very low	Very low	Low	Medium	Very low	High	High	Medium to high	Medium	Low
Part geometry	Complex	Simple	Simple	Simple	Simple	Some features	Some features	Simple	Simple	Some features
Part size	Very small to large	Small to medium	Small to very large	Large to very large	Small to medium	Small to very large	Small to very large	Small to very large	Small to large	Small to large
Tool cost	Very high	Low	Medium to high	Low to high	Low to high	Low	High	Low to medium	Medium	High
Cycle time	Seconds	Seconds	Seconds	Seconds to minutes	Seconds	Minutes	Minutes	Minutes	Minutes	Minutes
Labor	Automatic	Automatic	Automatic	Manual	Automatic	Manual	Automatic	Manual	Manual	Manual

Index

A

ABS. *See* Acrylonitrile butadiene styrene
Accumulators, blow molding, 52
Acrylic, 64
Acrylonitrile butadiene styrene (ABS)
    use of in extrusion, 36
    use of in rotational molding, 101
    use of in thermoforming, 70
Additives
    use of in extrusion materials, 36
    use of in injection molding, 15
    use of in thermoforming materials,
      72
Aluminum
    use of for blow mold tooling, 53
    use of for die materials in extrusion,
      35
    use of for molds, 14
    use of for reaction injection molding,
      82–83
    use of for rotational mold tooling,
      99–101
    use of for thermoforming mold
      fabrication, 69
Amorphous materials, use of in
      thermoforming, 70–72
Attachment methods
    blow molding design guidelines,
      55–56
    compression molding design
      guidelines, 114
    reaction injection molding design
      guidelines, 89
Automotive parts, use of structural
      reaction injection molding for,
      79

B

B.F. Goodrich, 7, 27, 108
Baekeland, Leo, 7, 108

Bakelite, 108
    use of in injection molding, 7
Barrels
    blow molding, 51
    extrusion, 29
    injection molding, 9
BASF, 7
Bayer AG, 79
Beryllium-copper, use of for blow mold
      tooling, 53
Bewley, H., 26
Blow molding
    equipment, 50–53
    history of, 49–50
    key process characteristics, 43t
    machines, 44–45f
    part design guidelines for, 55–56
    part identification, 56
    process overview, 43, 46–48
    process variations, 44
    tooling, 53
BMC. *See* Bulk molding compound
Boss
    compression molding design
      guidelines, 113, 114f
    injection molding design guidelines,
      17
    reaction injection molding design
      guidelines, 89
Bottle necks, blow molding design
      guidelines, 55
Bottles, blow molding of, 46–49
Breaker plates
    blow molding, 52
    extrusion, 32
Bulk molding compound (BMC), 112

C

Calibrator plates, 33–34f
Cap specifications, 54

143

Sheet extrusion, 25
Sheet metal, use of for rotational mold
      tooling, 99–101
Sheet molding compound (SMC), 112
Sheet processing, use of thermoforming
      for, 59–61
Shuttle configuration for rotational
      molding, 95
Silicone
    use of for reaction injection molding,
      82
    use of in compression molding, 109
Similar materials, case study, 40
Single screw extrusion, 30–31
SMC. *See* Sheet molding compound
Soft drink bottles, blow molding of,
      49–50
Soft-touch parts, 4, 8
Sprue, 14
SRIM. *See* Structural reaction injection
      molding
Stacking stage of thermoforming, 67
Stainless steel
    use of for blow mold tooling, 53
    use of for compression molding, 111
    use of for die materials in extrusion,
      35
    use of for injection molds, 14
Stationary platen, injection molding, 10
Steel
    use of for compression molding, 111
    use of for reaction injection molding,
      82–83
    use of for thermoforming mold
      fabrication, 69
Stereolithography
    use of for reaction injection molding
      molds, 83
    use of for thermoforming molds, 68
Streamline dies, 32–33f
Stretch blow molding (SBM), 47, 48f
Structural foam injection molding, 5–6
Structural reaction injection molding
      (SRIM), 79
Supply stage of thermoforming, 65

T

Terms, 117–140
Texture
    compression molding design
      guidelines, 112
    extrusion process design guidelines,
      38
    injection molding design guidelines,
      16
    thermoforming design guidelines,
      74
Thermoforming
    equipment used during stages of,
      65–68
    history of, 64
    identification of parts manufactured
      by, 74
    materials, 70–72
    part design guidelines, 73–74
    process key characteristics, 59t
    process overview, 59–61
    process variations, 61–64
    tooling, 68–70
Thermoplastics
    blow molding of, 50–53
    use of in extrusion processes, 36
Thermoset materials
    use of in compression molding, 107,
      112
    use of in extrusion processes, 36
Thinning, 73
Thread specifications, 54
Thrust bearing, 31
Tie bars, injection molding, 13
Tooling
    blow molding, 53
    compression molding, 111
    extrusion process, 35
    injection molding, 13–15
    rotational molding, 99–101
    thermoforming, 68–70
Toys, use of rotational molding for
      manufacture of, 97
Traffic construction barrels, case study,
      57–58
Traffic construction cones, case study,
      105
Transfer molding, 108